STOCHASTIC PROCESSES, ESTIMATION, AND CONTROL

STOCHASTIC PROCESSES, ESTIMATION, AND CONTROL

The Entropy Approach

GEORGE N. SARIDIS
Rensselaer Polytechnic Institute

A Wiley-Interscience Publication
JOHN WILEY & SONS, INC.
New York • Chichester • Brisbane • Toronto • Singapore

This text is printed on acid-free paper.

Library of Congress Cataloging in Publication Data:

Saridis, George N., 1931–
Stochastic processes, estimation, and control: the entropy approach / by George N. Saridis.
p. cm.
Includes bibliographical references and index.
ISBN 0-471-09756-X (acid-free)
1. Stochastic processes. 2. Control theory. 3. Entropy
I. Title
QA274.S37 1995
003′.76′0115—dc20 94-33109

Printed in the United States of America

10 9 8 7 6 5 4 3 2 1

TO
ELECTRA
PHAEDRA
and
YOULA

PREFACE

The concept of uncertainty is inherent to any kind of scientific investigation in the history of mankind. This is due to the fact that scientific tools, like mathematics, logic, reasoning, et cetera are simply used to model the real world by human standards, and represent only an approximation of the actual laws of nature.

Since the days of the Bernoulli brothers, scientists accounted for uncertainty by introducing probability measures to describe uncertain phenomena and express laws of nature. Furthermore, when prediction of uncertain performance of physical phenomena was required, for engineering, communication, and other analytic purposes, probabilistic models were proposed to account for the lack of complete information about the process under investigation.

It was not until the 1930s, when Kolmogorov presented his axiomatic probability theory, that a rigorous analytic device was created that could be reliably used to estimate, analyze, and provide tools for the design of analytic systems, useful for engineering purposes. Stochastic processes, state estimation, and optimal control of such dynamic systems, based on Kolmogorov's axiomatic theory of probability, are the result of scientific research of only the last 50 years. With the aid of the modern digital computers, their application made possible the miracles of space exploration, understanding and protecting the environment, and creating the information revolution of the late 20th century.

However, the underlying theory soon found analytic obstacles that limited its application to systems more general than linear. Various proposed approximations fell short of their goal, due to lack of understanding of the true meaning of the effect of the underlying uncertainties.

Entropy, which is introduced in thermodynamics, and was first used by Boltzmann to analyze uncertain phenomena, provides an excellent tool to measure and understand the behavior of systems under uncertainty. It has explained many of the discrepancies between deterministic and stochastic solutions of various problems in system theory, and has resolved their conceptual differences, such as Fel'dbaum's idea of active transmission of information. Entropy is used in this book to integrate the various approaches of analysis, design, and control of systems under uncertainty.

Because this book is intended as a text for a first-year graduate course, it presents a thorough review of stochastic processes, which represent the foundation of stochastic dynamic systems, an analytic discussion of parameter estimation based on least-squares techniques, an in-depth investigation of the estimation of the states of a stochastic linear and nonlinear dynamic system, and a modified derivation of the linear-quadratic-gaussian optimal-control problem. The method used for these derivations introduces first the discrete-time version of the problem and converts it to continuous time by simply taking the limits as the discrete-time interval goes to zero, thus avoiding the use of advanced mathematics like the Itô calculus. The new theory of approximation of the optimal solution for nonlinear stochastic systems is presented as a general engineering tool. Finally the whole subject of stochastic processes, estimation, and control is recast, using entropy as a measure.

References and comprehensive problems follow every chapter of the book.

This book would have been made impossible without the help of my students in my classes on stochastic estimation optimal and adaptive control at both Purdue University and Rensselaer Polytechnic Institute and especially my graduate students, to whom I express my gratitude. Special appreciation is due to Dr. Michael Mathioudakis and my wife Youla Saridis for proofreading the manuscript.

George N. Saridis

Loudonville, NY
November 1994

CONTENTS

STOCHASTIC PROCESSES, ESTIMATION, AND CONTROL

INTRODUCTION

0.1 MOTIVATION

System engineering deals mainly with the behavior of dynamics systems. The stochastic formulation of system engineering problems is a rather recent development. For example, only in the 1930s did Kolmogorov formulate his *axiomatic theory of probability*, which is the foundation of the theory of *stochastic processes*, and he and Wiener established *estimation theory* for dynamic systems in 1941. Of course the Bernoulli brothers, Bayes, Gauss, and others have been publishing articles on *probability theory* and the least-squares method of parameter estimation since the early 18th century [Sorenson (1980)].

It was later, in the 1950s and 1960s, that *state estimation* was finalized by Kalman and Bucy (1961), *stochastic optimal control* was introduced in the system literature [Fel'dbaum (1965), Tou and Joseph (1961), Meditch (1969), Pontryagin et al. (1962), etc.], and *parameter identification* was made legitimate among system scientists [Lee (1964), Sage and Melsa (1971), Saridis (1977), etc.]. It was not, however, until the 1980s that the theories underlying the problems were well understood and clearly explained.

The theory of *stochastic processes*, *estimation*, *and control* is based on the concept of *uncertainty*, which is imbedded in human reasoning. Because we are always uncertain about the precision with which we describe the laws of nature, probability theory maps the various system problems we are trying to solve to a probability space of the measure of uncertainty, in which we can utilize the ingenious mathematical methods scientists have developed.

It is, then, only natural to search for a measure of uncertainty, to find a common measure of performance for all the system problems we are trying to solve as engineers. This is especially true with problems that deal with the performance and design of *Intelligent Machines*, like modern robots, equipped with autonomous and semiautonomous sensory, control, planning, and manipulative systems, all working in unison. Because most laws of nature were derived through energy concepts, it is not unreasonable to look for an energy measure as the *unifying measure of performance* for such engineering systems. *Entropy* was introduced by Clausius in 1867, to describe the form of thermal energy converted according to the Second Law of Thermodynamics. Later Boltzmann (1872) and Shannon (1963) used entropy to ascribe a measure of uncertainty. After a considerable amount of research [Saridis (1985)], it was concluded that entropys is the proper measure to describe the performance of systems consisting of components from diversified disciplines. This is accomplished by introducing the concept of uncertainty in the design and the measure of performance of these systems.

0.2 BOOK ORGANIZATION

The purpose of this book is to demonstrate the reformulation of stochastic estimation and control problems with entropy as a measure of performance. In order to accomplish this goal, and provide a text on the subject for the practicing engineer, it is necessary to present the basic topics on stochastic processes, parameter estimation, and optimal control to the reader. The approach followed is purely in the time domain, because nonlinear estimation and control cannot be presented thoroughly in the frequency domain. However, both discrete-time and continuous-time representations are considered in the above order because they result in a simplified description of the material, without the need for sophisticated calculus methods not familiar to the unskilled reader.

The organization of the book is as follows: Chapter 1 presents an introduction of the basics on stochastic processes from the time-domain point of view; Chapter 2 discusses the fundamentals of parameter estimation both from the deterministic and the probabilistic point of view; Chapter 3 gives a rigorous account of the problem of state estimation both in the continuous- and discrete-time cases, covering linear as well as nonlinear filtering; Chapter 4 derives the most recent results on stochastic optimal control with emphasis on the linear-quadratic-gaussian problem, in the continuous- and discrete-time cases; Chapter 5 presents an original derivation of an approximation theory for continuous-time nonlinear stochastic optimal control as the only feasible approach for an engineering solution of this problem; finally, Chapter 6 presents the reformulation of the problems of state estimation and optimal control, using entropy as a measure of performance, with special emphasis on the dual and the adaptive control problems. The question of the active transmission of information, introduced by Fel'dbaum, is then resolved. Each chapter is followed by a set of training exercises, and a pertinent reference list.

REFERENCES

Boltzmann, L., "Further Studies on Thermal Equilibrium Between Gas Molecules" *Wien Ber.*, **66**, 275 (1872).

Fel'dbaum, A. A., *Optimal Control Systems*, Academic Press, New York, 1965.

Kalman, R. E., and Bucy R., "New Results in Linear Filtering and Prediction Theory" *ASME Trans. J. Basic Eng.*, 83–95 (1961).

Lee, R. C. K., *Optimal Identification, Estimation, and Control*, MIT Press, Cambridge, MA, 1964.

Meditch, J., *Stochastic Linear Estimation and Control*, McGraw-Hill, New York, 1969.

Pontryagin, L. S., Boltyanskii, V. G., Gamkrelidge, R. V., and Mishchenko, E. F., *The Mathematical Theory of Optimal Processes*, John Wiley and Sons, New York, 1962.

Sage, A. P., and Melsa, J. L., *System Identification*, Academic Press, New York, 1971.

Saridis, G. N., *Self-Organizing Controls of Stochastic Systems*, Marcel Dekker, New York, 1977.

Saridis, G. N., "An Integrated Theory of Intelligent Machines, by Expressing the Control Performance as an Entropy," CTAT Vol. 1, No. 2 pp. 125–138 1985.

Shannon, C., and Weaver, W., *The Mathematical Theory of Communications*, Illini Books, Urbana, IL, 1963.

Tou, J., and Joseph, P., "On Linear Control Theory" *AIEE Trans.*, **80**(19) (1961).

Sorenson, H. W., *Parameter Estimation: Principles and Problems*, Marcel Dekker, New York, 1980.

Wonham, W. M., "Random Differential Equations in Control Theory," in Bharucha-Reid, A.T. (Ed.), *Probabilistic Methods in Applied Mathematics*, Academic Press, New York, 1970.

1

STOCHASTIC PROCESSES

1.1 DEFINITIONS

A *Stochastic process* is a collection or an *ensemble* of time functions $X(\zeta, t)$ of random variables, corresponding to an infinitely countable set of experiments ζ, with an associated probability description, for example $p(x, t)$. See Figure 1.1. $X(t_1)$ is a random variable with probability $p(x, t_1)$; $X(t_2)$ is a random variable with probability $p(x, t_2)$.

EXAMPLE An Electromotive force generated by a voltage generator ζ is randomly selected as in Figure 1.2.

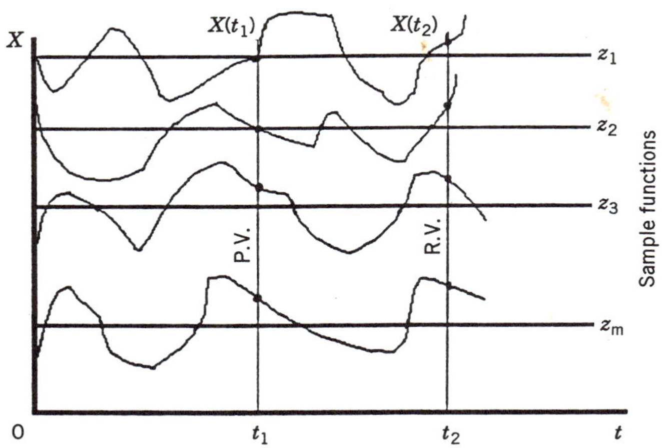

Fig. 1.1 Ensemble of stochastic processes

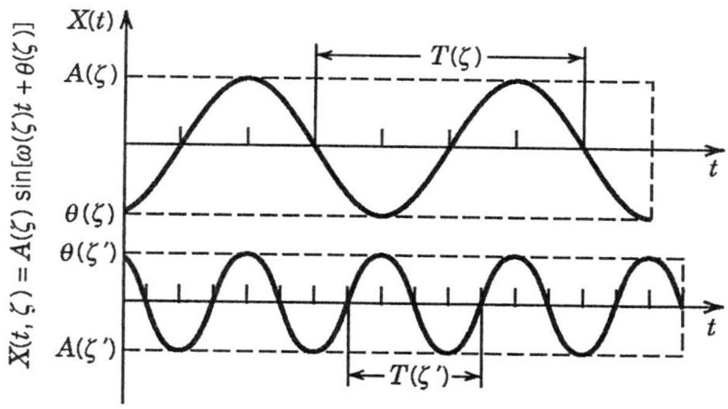

Fig. 1.2 Randomly generated Electromotive Force

1.2 STATISTICS OF STOCHASTIC PROCESSES

Because t_i defines a random variable over the ensemble every time, various statistics may be defined among these random variables. We shall consider the first- and second-order statistics only.

First-Order Statistics. First-order statistics treat only one random variable $X(t)$, and are described by a probability distribution function $P(x, t)$ and its density $p(x, t) = dP/dt$ for the continuous-time case, and by the distribution function $P(x_i, t)$ for the discrete-time case.

Second-Order Statistics. Second-order statistics treat two random variables at two separate time instances t_1 and t_2, $X(t_1)$ and $X(t_2)$. For the continuous-time case the following probability functions are defined, for $x(t_1) = x_1$, $x(t_2) = x_2$:

Joint distribution: $P(x(t_1), x(t_2); t_1, t_2)$
Joint density: $p(x(t_1), x(t_2); t_1, t_2) = dP/dx_1\, dx_2$
Marginal density: $p(x_2, t_2) = \int_{-\infty}^{\infty} p(x_1, x_2; t_1, t_2)\, dx_1$
Conditional density: $p(x_1, t_1/x_2, t_2) = p(x_1, x_2; t_1, t_2)/p(x_2, t_2)$

First- and second-order moments are defined as follows:

Mean value: $\bar{x}(t) = E\{X(t)\} = \int_{-\infty}^{\infty} xp(x, t)\, dx$

Autocorrelation function:

$$R_{xx}(t_1, t_2) = \int_{-\infty}^{\infty}\int_{-\infty}^{\infty} x_1 x_2\, p(x_1, x_2; t_1, t_2)\, dx_1\, dx_2$$

Autocovariance: $C_{xx}(t_1, t_2) = E\{[x(t_1) - \bar{x}(t_1)][x(t_2) - \bar{x}(t_2)]$
$= R_{xx}(t_1, t_2) - \bar{x}(t_1)\bar{x}(t_2)$

Covariance: $C_{xx}(t, t) = \sigma_x^2.$

1.3. TRANSFORMATIONS OF STOCHASTIC PROCESSES

1.3.1 Functions of One Random Variable (Appendix B)

Fundamental Theorem 1.1 To find the density $p_Y(y)$ of the function of a random variable x, with $p_X(x)$ defined:

$$y = g(x), \tag{1.1}$$

with $x_1 \cdots x_n$ the real roots of Eq. (1), satisfying

$$y = g(x_1) = \cdots = g(x_n) \Rightarrow x_i = g(y)^{-1} = f(y); \tag{1.2}$$

then

$$p_Y(y) = p_X(x_1)/|dg(x_1)/dx| + \cdots + p_X(x_n)/|dg(x_n)/dx|. \tag{1.3}$$

1.3.2 Stochastic Processes

Consider the transformation T:

$$y(t) = T[x(t)]. \tag{1.4}$$

A deterministic transformation is defined if

$$x(t, \zeta_1) = x(t, \zeta_2) \Rightarrow y(t, \zeta_1) = y(t, \zeta_2).$$

A nondeterministic transformation is defined if

$$x(t, \zeta_1) = x(t, \zeta_2) \Rightarrow y(t, \zeta_1) \neq y(t, \zeta_2).$$

Statistics of $y(t)$ are defined as statistics of $x(t)$:

$$p_Y(y, t) = \sum_i p_x(x_i, t)/|dg(x_i, t)/dx|. \tag{1.5}$$

Special Case. A time-invariant system with no memory:

$$y(t) = g[x(t)], \tag{1.6}$$

for which,

$$y(t) = g[x_i(t)], \qquad i = 1 \cdots n. \tag{1.7}$$

EXAMPLE

$$y(t) = g(x) = x(t)^2$$

with $p_X(x)$ given:

$$x_1(t) = \sqrt{y(t)}, \qquad x_2(t) = \sqrt{y(t)}, \qquad |dy/dx| = 2\sqrt{y(t)};$$

then

$$p_Y(y) = \left[p_X(\sqrt{y}) + p_X(-\sqrt{y})\right]/2\sqrt{y}.$$

If $y(t) < 0$, then $p_Y(y) \equiv 0$.

Similarly, joint densities may be found via stochastic transformations. Moments can be obtained as well for

$$y(t) = g[x(t)],$$

$$E\{y(t)\} = \int_{-\infty}^{\infty} g(x(t))p_X(x,t)\,dx$$

$$R_{xx}(t_1,t_2) = \int_{-\infty}^{\infty}\int_{-\infty}^{\infty} g(x_1)g(x_2)p(x_1,x_2;t_1,t_2)\,dx_1,dx_2, \tag{1.8}$$

et cetera.

1.3.3 Sample Statistics and Sample Moments

Sample statistics time averages over the sample functions of the process

$$\langle X^n \rangle = \lim_{T\to\infty} (1/2T)\int_{-T}^{T} X(t)^n\,dt. \tag{1.9}$$

For the equivalent discrete-time ensemble and sample moments see Appendix B.

1.4 TYPES OF STOCHASTIC PROCESSES

1. Continuous or discrete
2. Deterministic or nondeterministic
3. Stationary or nonstationary
4. Ergodic or nonergodic

1.4.1 Continuous or Discrete Stochastic Processes

A *continuous* stochastic process is one with continuous probability density function (no delta functions).

A *discrete* stochastic process is one with random variables assuming only certain isolated values (see Figure 1.3) where the probability distribution function is discretely defined. The density function assumes delta functions at these points. Mixed processes have both continuous and discrete components.

1.4.2 Deterministic or Nondeterministic Stochastic Processes

A *deterministic* stochastic process is one wherein the future values are predictable once we know one point of a sample function:

$$x(t) = A\cos(\omega t + \theta),$$

where θ is a random variable over the ensemble. If the random variable is unpredictable from observed values it is *nondeterministic*.

For deterministic stochastic processes, sample statistics, for example, time averages and moments describe the process. Measurements of process parameters are defined in Section 1.3.3.

1.4.3 Stationary or Nonstationary Stochastic Processes

If all marginal and joint density functions of a stochastic process do not depend on the choice of time origin the process is *stationary*. If not it is *nonstationary*. Deterministic processes are usually stationary under special conditions.

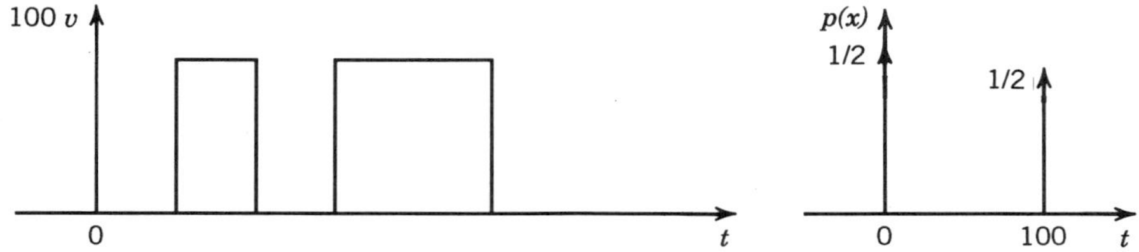

Fig. 1.3 Randomly opening switch

1.4.4 Ergodic or Nonergodic Stochastic Processes

Ergodicity is defined when the ensemble moments are equal to the sample moments of a stochastic process:

$$E\{X^n\} = \int_{-\infty}^{\infty} x^n p(x)\,dx = \lim_{T\to\infty} 1/2T \int_{-T}^{T} X(t)^n\,dt. \tag{1.10}$$

Ergodic stochastic processes are always stationary. The converse is not always true.

1.5 AN ILLUSTRATIVE CASE STUDY

Consider the voltage generation equation of Figure 1.4 as a case for the study of various properties of a stochastic process:

$$X(t) = Y\cos(\omega t + \theta), \tag{1.11}$$

where Y is the voltage amplitude, a random variable with $p_Y(y)$ density function, ω is a constant frequency, and θ is the unknown phase angle, an independent random variable uniformly distributed over $[0, k\pi]$, $k = 1, 2$. It is obvious from the previous definitions that (1.11) represents a continuous deterministic stochastic process.

1.5.1 Derivation of Density Function

The density function $p_X(t)$ can be obtained in two steps by considering

$$X = Y \cdot Z, \qquad \text{where } Z = \cos(\omega t + \theta).$$

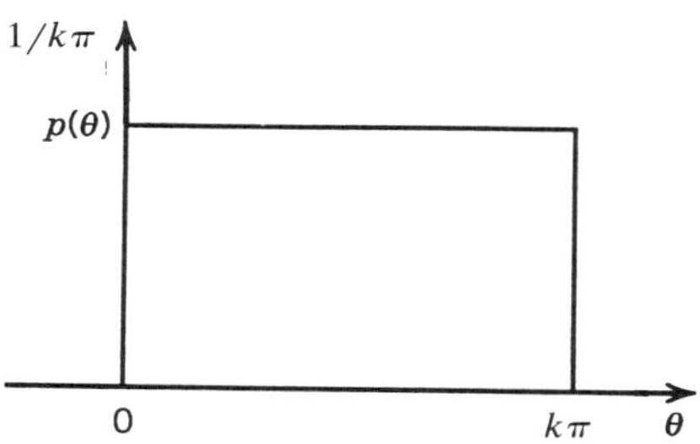

Fig. 1.4 Uniform density

(a) For $k = 2$, θ assumes the following values, solutions of the equations,

$$z = \cos(\omega t + \theta_1) = \cos(\omega t + \theta_2),$$

$$\theta_1 = -\omega t + \cos^{-1} z, \qquad \partial\theta_1/\partial z = -1/\sqrt{1 - z^2},$$

$$\theta_2 = 2\pi - \omega t + \cos^{-1} z, \qquad \partial\theta_2/\partial z = 1/\sqrt{1 - z^2}.$$

Then the density of Z is given by the transformation theorem:

$$\begin{aligned} p_Z(z) &= \left|1/\sqrt{1 - z^2}\right|\left[p_\Theta(-\omega t + \cos^{-1} z) + p_\Theta(2\pi - \omega t - \cos^{-1} z)\right] \\ &= \left|1/\sqrt{1 - z^2}\right|\left[1/2\pi + 1/2\pi\right] = 1/\pi\sqrt{1 - z^2} \quad -1 < z < 1. \end{aligned} \tag{1.12}$$

(b) The joint distribution of

$$X = Y \cdot Z \text{ over } D_X = \left[\begin{array}{l} -1 < z < 1, \\ -\infty < y < \infty \end{array}\right.$$

is given by

$$P_X(x) = \text{Prob}(X \leq yz) = \iint_{D_X} p_Y(y) p_Z(z)\, dy\, dz.$$

However,

$$D_X = \{x \leq yz\} = \begin{cases} x/z \leq y, & 0 \leq z < 1, \\ x/z \geq y, & -1 < z \leq 0. \end{cases}$$

Therefore, adjusting the limits of integration,

$$P_X(x) = \int_{-1}^{0} p_Z(z) \int_{-\infty}^{x/z} p_Y(y)\, dy\, dz + \int_0^1 p_Z(z) \int_{x/z}^{\infty} p_Y(y)\, dy\, dz.$$

Then the density function is

$$\begin{aligned} p_X(x) = dP_X(x)/dx &= \int_{-1}^{0} p_Z(z) p_Y(x/z)/z\, dz - \int_0^1 p_Z(z) p_Y(x/z) z\, dz \\ &= \int_{-1}^{1} p_Z(z) p_Y(x/z)|z|\, dz = 1/\pi \int_{-1}^{1} p_Y(x/z) \Big/ \left[|z|\sqrt{1 - z^2}\right] dz. \end{aligned} \tag{1.13}$$

If $p_Y(y)$ is independent of time $p_X(x)$ is also independent of time and $X(t)$ is *stationary*.

From the above derivation and after substitution of the appropriate densities $p_Z(z)$, $p_Y(y)$, it becomes obvious that obtaining the probability function for a rather simple case is a major task. Therefore, it is easier to study the properties of a stochastic process through its moments. Guided by the *Gaussian density*, which is completely described by its *mean* and *variance*, one is inclined to utilize the first and second moments of a stochastic process only, thus exploring its properties in the *wide sense*.

1.5.2 Stationarity in the Wide Sense

The ensemble mean of the stochastic process (1.11) is given by

$$\begin{aligned} E\{X(t)\} &= E\{Y\cos(\omega t+\theta)\} = E\{Y\}E\{\cos(\omega t+\theta)\} \\ &= \bar{Y}\int_0^{2\pi}\cos(\omega t+\theta)\,d\theta = (\bar{Y}/2\pi)[\sin(\omega t+2\pi)-\sin(\omega t)] \\ &= 0, \qquad k=2 \end{aligned} \tag{1.14}$$

or

$$\begin{aligned} &= \bar{Y}\int_0^{2\pi}\cos(\omega t+\theta)\,d\theta = (\bar{Y}/2\pi)[\sin(\omega t+\pi)-\sin(\omega t)] \\ &= \bar{Y}/\pi\sin(\omega t), \qquad k=1. \end{aligned}$$

It is obvious from the above that for $k = 1$, the stochastic process (1.11), is *not* stationary, but for $k = 2$ it is a candidate for stationarity in the wide sense depending on the second moment.

The second moment of (1.11), for $k = 2$, is given by

$$\begin{aligned} E\{X^2\} &= E\{Y^2\}E\{\cos^2(\omega t+\theta)\} = \bar{Y}^2E\{1/2[1+\cos(2\omega t+2\theta)]\} \\ &= \bar{Y}^2/2\left[1+\int_0^{2\pi}1/2\pi\cos(2\omega t+\theta)\,d\theta\right] \\ &= \bar{Y}^2/2\left[1+1/4\pi\int_{2\omega t}^{4\pi+2\omega t}\cos u\,du\right] = Y^2/2. \end{aligned} \tag{1.15}$$

Because the first and second moments of (1.11) are independent of time the stochastic process is stationary in the wide sense.

1.5.3 Ergodicity in the Wide Sense

In order to study the ergodicity in the wide sense, the sample first and second moments should be derived. The ith sample mean is

$$\langle X_i\rangle = \lim_{T\to\infty}(1/2T)\int_{-T}^{T}X_i(t)\,dt = Y_i\lim_{T\to\infty}(1/2T)\int_{-T}^{T}\cos(\omega t+\theta_i)\,dt = 0, \tag{1.16}$$

where Y_i and θ_i are the amplitude and phase angle of the ith sample function. Therefore, because

$$\langle X_i \rangle = E\{X(t)\} = 0, \qquad \forall\, i,$$

the process is a candidate for ergodic in the wide sense.

The second moment of the ith sample is given by

$$\begin{aligned}
\langle X^2 \rangle &= \lim_{T\to\infty} (1/2T)\int_{-T}^{T} Y_i^2 \cos(\omega t + \theta_i)\, dt \\
&= Y_i^2 \lim_{T\to\infty} (1/4T)\int_{-T}^{T} [1 + \cos 2(\omega t + \theta_i)]\, dt \\
&= Y_i^2 \lim_{T\to\infty} 1/8T\omega \int_{-2\omega T+\theta i}^{2\omega T+\theta i} [1 + \cos 2(\omega t + \theta_i)]\, d(\omega t + \theta_i) \\
&= Y_i^2 \lim_{T\to\infty} 1/8T\omega[4\omega T + \sin(2\omega T + \theta_i)] - \sin(-2\omega T + \theta_i) \\
&= Y_i^2\Big[\tfrac{1}{2} + 1/8\omega T \lim_{T\to\infty} [\sin(2\omega T + \theta_i) + \sin(2\omega T - \theta_i)]\Big] \\
&= Y_i^2\Big[\tfrac{1}{2} + 1/4\omega T \lim_{T\to\infty} [\sin 2\omega T \cos 2\theta_i]\Big] \\
&= Y_i^2\Big[\tfrac{1}{2} + \tfrac{1}{2}\cos 2\theta_i \lim_{T\to\infty} [\sin 2\omega T/2\omega T]\Big] = Y_i^2/2, \qquad \omega \neq 0.
\end{aligned} \tag{1.17}$$

Therefore, the process (1.11) is nonergodic, because

$$\langle X^2 \rangle \neq E\{X^2\}. \tag{1.18}$$

The above examples have demonstrated the various properties of the stochastic processes.

1.6 STATIONARITY AND ERGODICITY

In a previous section the density function of a transformed stochastic process was presented. This was also illustrated in the previous case study when a stochastic process was generated by two random variables. In that case, stationarity and ergodicity of the transformed stochastic process was established or disputed, depending on properties of the time-varying transformation utilized and the random variables involved. The general rule for a stationary process $X(t)$ to be transformed to another stationary process $Y(t)$, is that the transformation is independent of time:

$$Y(t) = g[X(t)] \rightarrow p_Y(y) = \sum_i p_X(x_i)/|dg(x_i)/dx|. \tag{1.19}$$

In order to guarantee stationarity in the strict sense, in addition to condition (1.19), all the joint moments must also be independent of time. Let

$$Y(t_i) = Y_i = g[X(t_i)] = g[X_i], \qquad i = 1 \cdots k \quad \text{and}$$
$$X_i^j = g^{-1}[Y_i], \qquad j = 1 \cdots n,$$

and the Jacobian of the transformation, $J = [dg(x_i^j)/dx_i]$; then the k-joint density function is

$$p_Y(y_1 \cdots y_k) = \sum_j^n p_X(x_1^j \cdots x_k^j) \Big/ \Big[|dg(x_1^j)/dx_1| \cdots |dg(x_k^j)/dx_k| \Big]. \tag{1.20}$$

For *stationarity in the wide sense*, on the other hand, which is easier to test, only the second-order ($k = 2$) joint density is sufficient.

EXAMPLE Consider the transformation presented by Eq. (1.7), for which the density function has been found already:

$$Y(t) = X(t)^2 \Rightarrow p_Y(y) \begin{cases} = \left[p_X(\sqrt{y}) + p_X(-\sqrt{y}) \right] \big/ 2\sqrt{y}, & y(t) > 0, \\ \equiv 0, & y(t) < 0. \end{cases}$$

The joint density function $p_Y(y_1, y_2)$ is obtained, for $i, j = 1, 2$, by

$$y_1 = x_1^2, \qquad y_2 = x_2^2;$$
$$\Rightarrow x_i^j = \left(\sqrt{y_1}, \sqrt{y_2}\right), \left(-\sqrt{y_1}, \sqrt{y_2}\right), \left(\sqrt{y_1}, -\sqrt{y_2}\right), \left(-\sqrt{y_1}, -\sqrt{y_2}\right).$$

The absolute value of the degenerate Jacobian is

$$|J| = 4|x_1 x_2| = 4\sqrt{y_1}\sqrt{y_2},$$

and the joint density function is

$$p_Y(y_1, y_2) = \left[p_X\left(\sqrt{y_1}, \sqrt{y_2}\right) + p_X\left(-\sqrt{y_1}, \sqrt{y_2}\right) + p_X\left(\sqrt{y_1}, -\sqrt{y_2}\right) + p_X\left(-\sqrt{y_1}, -\sqrt{y_2}\right) \right] \Big/ 4\sqrt{y_1}\sqrt{y_2}.$$

It is obvious that the process is stationary in the wide sense. However, the converse is not always true, as was demonstrated in the previous case study.

The test for *ergodicity in the wide sense* is more complicated, requiring the calculation of the first and second moments of the ensemble and the sample functions every time, as demonstrated in the case study.

1.7 APPLICATIONS

Various processes modeled as applications of stochastic processes will be presented in the following.

1.7.1 Random Binary Transmission

Consider the processes of transmitting a binary electric signal [Figure 1.5(a)] at the outcome of a heads- or -tails experiment:

$$X(t) = \begin{cases} 1, & p = \frac{1}{2} \text{ heads}, \quad (n-1)T < t < nT, \\ -1, & q = \frac{1}{2} \text{ tails}. \end{cases} \tag{1.21}$$

The first two moments are

$$E\{X(t)\} = 0, \qquad E\{X(t)^2\} = 1, \qquad R(t_1, t_2) = E\{x(t_1), x(t_2)\} = 1,$$
$$(n-1)T < t_1, t_2 < nT. \tag{1.22}$$

1.7.2 Random Binary Reception

The random binary reception signal [Figure 1.5(b)] is the random binary transmission signal delayed by a random time, uniformly distributed over the time interval $[0, T]$, independent of the transmission process:

$$Y(t) = X(t-e), \qquad \text{with } E\{Y(t)\} = E\{E\{X(t-e)/e\}\} = 0, \tag{1.23}$$

because $E\{X(t-e)/e\} = 0$.

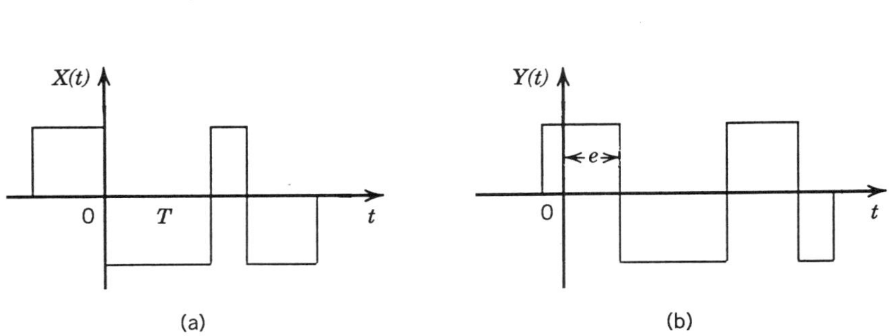

Fig. 1.5 (a) Binary transmission. (b) Binary reception

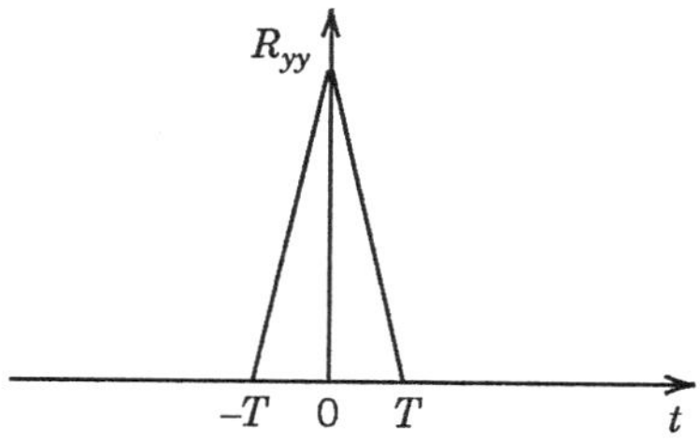

Fig. 1.6 Autocorrelation function

The autocorrelation function (Figure 1.6) is claimed to be

$$R_{yy}(t_1 t_2) = E\{Y(t_1)Y(t_2)\} = \begin{cases} 0, & \text{if } |t_1 - t_2| > T, \\ 1 - |t_1 - t_2|/T, & \text{if } |t_1 - t_2| < T. \end{cases} \tag{1.24}$$

Proof: If $|t_1 - t_2| > T$ for $e < T$, $Y(t_1)$ and $Y(t_2)$ are in different pulses and

$$R_{yy}(t_1, t_2) = E\{Y(t_1)Y(t_2)/e\} = 0.$$

If $|t_1 - t_2| < T$ and $t_1 = nT$ implies that $e < T - |t_1 - t_2|$, and t_1, t_2 are in the same pulse, then

$$R_{yy}(t_1, t_2) = E\{Y(t_1)Y(t_2)/e\} = \begin{cases} 1, & e < T - |t_1 - t_2|, \\ 0, & \text{elsewhere}, \end{cases}$$

and

$$R_{yy}(t_1, t_2) = E\{Y(t_1)Y(t_2)\} = 1 \cdot P\{e < T - |t_1 - t_2|\} = 1 - |t_1 - t_2|/T.$$

Q.E.D.

1.7.3 Markov Process

A Markov process is the process wherein, for $t_1 > t_2 > \cdots > t_n$,

$$\begin{aligned} &\text{Prob}\{X(t_1) \le x_1/X(t_2) = x_2 \cdots X(t_n) = x_n\} \\ &\quad = \text{Prob}\{X(t_1) \le x_1/X(t_2) = x_2\}. \end{aligned}$$

In this case x_2 is a sufficient statistic for the process.

1.7.4 Poisson Process

The Poisson process is the process wherein the number $n(t_1, t_2)$ of $t_i \in (t_1, t_2)$ of length $t = t_1 - t_2$ has the probability distribution with the Poisson parameter λt:

$$P\{n(t_1, t_2) = k\} = e^{-\lambda t}(\lambda t)^k/k!. \tag{1.25}$$

If $t_1 > t_2 > t_3 > t_4$, then $n(t_1, t_2)$ is independent of $n(t_3, t_4)$.

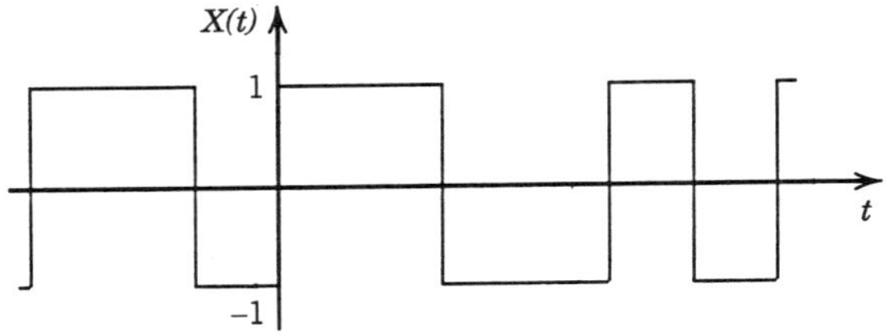

Fig. 1.7 Telegraph signal

The resulting stochastic process $X(t)$ is discrete state, and has the properties

$$X(t) = n(0, t),$$

$$E\{X(t)\} = \eta(t) = \lambda t, \tag{1.26}$$

$$R_{xx}(t, t_2) = \begin{cases} \lambda t_2 + \lambda t_1 t_2, & t_1 \geq t_2, \\ \lambda t_1 + \lambda^2 t_1 t_2, & t_1 \leq t_2, \end{cases}$$

$$E\{X(t)^2\} = \lambda t + \lambda^2 t_2. \tag{1.27}$$

1.7.5 Telegraph Signal

A telegraph signal (Figure 1.7) is a stochastic process $[-1, 1]$ with probability distribution

$$P\{X(t) = 1\} = e^{-\lambda} \cosh(\lambda t),$$

$$P\{X(t) = -1\} = e^{-\lambda} \sinh(\lambda t),$$

and

$$R_{xx}(t_1, t_2) = \exp\{-2\lambda|t_1 - t_2|\}. \tag{1.28}$$

1.7.6 Random Walk (Discrete-Time Brownian Motion)

Consider the process generated by making a step to the right when the outcome of a coin-tossing experiment is heads and a step to the left when it is tails, at every T seconds. The position t s after the first toss is $X(t)$, which gave the process the name *random walk*. See Figure 1.8.

After n tossings of the coin k heads appeared at $t = nT$, equivalent to k steps (s) to the right, and $n - k$ steps to the left, or

$$X(nT) = ks - (n - k)s = (2k - n)s = rs \quad \text{and} \quad k = (r + n)/2,$$

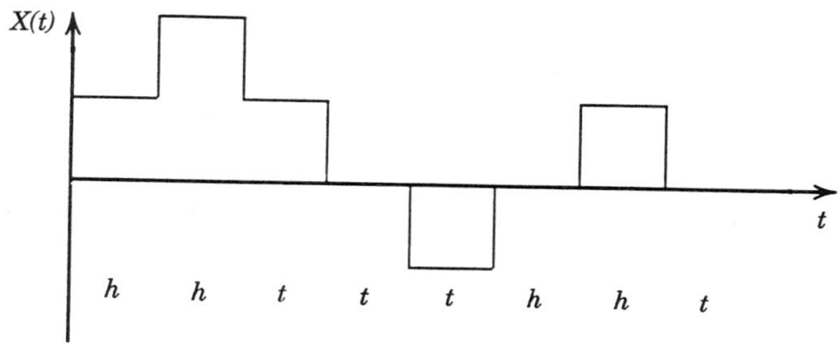

Fig. 1.8 Random walk

which implies that

$$\text{Prob}\{X(nT) = rs\} = \text{prob}\{(r+n)/2 \text{ heads}\} = \binom{n}{(n+r)/2} 1/2^n. \tag{1.29}$$

Now, to show that

$$E\{X(nT)\} = 0 \qquad E\{X(nT)^2\} = ns^2,$$

assume that X_i is the ith step of the process:

$$\text{Prob}\{X_i = s\} = \tfrac{1}{2}, \qquad \text{Prob}\{X_i = -s\} = \tfrac{1}{2} \Rightarrow E\{X_i\} = 0, \qquad E\{X_i^2\} = s^2.$$

But

$$X(nT) = \sum_i^n X_i \Rightarrow E\{X(nT)\} = 0, \quad \text{and} \quad E\{X(nT)^2\} = \sum_i^n E\{X_i^2\} = ns^2.$$

Asymptotically as $n \to \infty$, by the DeMoivre–Laplace theorem:

$$\text{Prob}\{k \text{ heads}\} \approx 1/\sqrt{2\pi npq}\, \exp\left[-(k-np)^2/2npq\right].$$

Because $p = q = \frac{1}{2}$ and $k = \frac{1}{2}(n+r)$,

$$\text{Prob}\{X(nT) = rs\} \approx 1/\sqrt{n\pi/2}\, \exp\left[-r^2/2n\right], \qquad n \gg 1,$$

where r is of order $\sqrt{n}$, and

$$\text{Prob}\{X(nT) \leq rs\} \approx \tfrac{1}{2} + \text{erf}\left[r/\sqrt{n}\right]. \tag{1.30}$$

For $n_1 > n_2 > n_3 > n_4$, the increments $X(n_1T) - X(n_2T)$ and $X(n_3T) - X(n_4T)$ are independent of each other. This process is extremely important because its variance is proportional to n and not n^2 as expected.

1.7.7 Wiener–Lévy Process (Brownian Motion)

Brownian motion was discovered in random particle motion in physics. It may be considered as the continuous-time version of the random-walk process, as $t \to 0$. Let

$$t = nT, \qquad X(nT) \to X(t), \qquad (n-1)T < t \leq nT. \tag{1.31}$$

Then, based on the previous derivations for discrete time;

$$E\{X(t)\} = 0, \qquad E\{X(t)^2\} = ns^2 = (t/T)s^2:$$

as $s \to 0$ and $T \to 0$, $E\{X(t)^2\} > 0$, if $s \sim \sqrt{T}$ or $s^2 = \alpha T$, for some α constant. Define the process $W(t)$;

$$W(t) = \lim_{T \to 0} X(t).$$

Then

$$E\{W(t)\} = 0, \qquad E\{W(t)^2\} = \alpha t. \tag{1.32}$$

If $w = rs$, $t = nT$, then $r/\sqrt{n} = w/s\sqrt{t/T} = w/\sqrt{\alpha t}$, and

$$\begin{aligned} \mathrm{Prob}\{W(t) \leq w\} &= \tfrac{1}{2} + \mathrm{erf}\left[w/\sqrt{\alpha t}\right], \\ p(w,t) &= 1/\sqrt{2\pi\alpha t}\, \exp\left[-w^2/2\alpha t\right], \\ R_{xx}(t_1, t_2) &= \begin{cases} \alpha t_2 t_1 \geq t_2, \\ \alpha t_1 t_1 \leq t_2. \end{cases} \end{aligned} \tag{1.33}$$

Proof: If $t_1 \geq t_2$, then $w(t_1) - w(t_2)$ is independent of $w(t_2) - w(0)$, as claimed in the discrete-time case. Hence,

$$\begin{aligned} E\{[w(t_1) - w(t_2)]w(t_2)\} &= E\{w(t_1)w(t_2)\} - E\{w(t_2)^2\} \\ &= E\{w(t_1)w(t_2)\} - \alpha t_2 = 0, \end{aligned}$$

or

$$R_{xx}(t_1, t_2) = E\{w(t_1)w(t_2)\} = \alpha t_2, \qquad \text{for } t_1 \geq t_2.$$

Similarly,

$$R_{xx}(t_1, t_2) = \alpha t_1, \qquad \text{for } t_1 \leq t_2. \qquad \text{Q.E.D.}$$

The above result is of major importance for the definition of stochastic differentiation of stochastic processes with Gaussian white noise as derivatives. It explains why second-order differentials in x, being of the same order as the first-order time differentials, must be included in the formula. The problem will be further discussed in the next section.

1.8 STOCHASTIC CONVERGENCE, DIFFERENTIATION, AND INTEGRATION

Differentiation and integration of stochastic processes cannot be defined in the Riemann sense, because they involve convergence of random sequences that may contain values at infinity with infinitesimal probability. A review of convergence of random sequences and stochastic continuity will precede the definition of stochastic differentials and integrals, in the following.

1.8.1 Convergence of Random Sequences

Consider the sequence of random variables $X = \{x_1, x_2 \cdots x_n\}$. It actually constitutes a random process, if the indices are understood to represent increasing time instances. Convergence of the sequence X may be defined in various ways depending on the measure used to express it. Some of these definitions are given hereafter.

***Convergence with Probability 1* (w.p. *1*).**

$$\text{Prob}\Big[\lim_{n \to \infty} \|x_n - x\| = 0\Big] = 1. \tag{1.34}$$

Convergence in the Mean-Square Sense (in* m.s.*).

$$\lim_{n \to \infty} E\{\|x_n - x\|^2\} = 0. \tag{1.35}$$

Convergence in Probability (in* p.*).

$$\lim_{n \to \infty} \text{Prob}\{\|x_n - x\| > \epsilon\} = 0. \tag{1.36}$$

Convergence w.p. 1 and in m.s. are the strongest measures of convergence. However, one does not imply the other; they just overlap each other. Both of them, though, imply the weaker convergence in p. The proof of this is usually established by the following useful relations, discussed in Appendix B.

Chebychev Inequality.

$$\operatorname{Prob}\{\|x_n - x\| > \epsilon\} \leq E\{\|x_n - x\|^2\}/\epsilon^2. \tag{1.37}$$

Law of Large Numbers. This was originally developed as a corollary of the DeMoivre–Laplace theorem:

$$\lim_{n\to\infty} \operatorname{Prob}\left\{\left|(1/n)\sum_i^n x_i - p\right| \leq \epsilon\right\} = 1, \tag{1.38}$$

where p is the probability of successful events ($x_i = 1$).

Strong Law of Large Numbers. This is a stronger version of the previous law (Borel):

$$\operatorname{Prob}\left\{\lim_{n\to\infty}\left|(1/n)\sum_i^n x_i - p\right| \leq \epsilon\right\} = 1. \tag{1.39}$$

The **central limit theorem**, responsible for our preference for Gaussian distributions, establishes the asymptotic behavior of any random sequence to be normal. Let

$$E\{x_i\} = \eta_i, \qquad \sigma_{xi}^2 = \sigma_i^2 \quad \text{and} \quad x = \sum_i x_i, \qquad \eta = \sum_i \eta_i, \qquad \sigma^2 = \sum_i \sigma_i^2;$$

$$p(x) \xrightarrow[n\to\infty]{} 1/\sigma\sqrt{2\pi}\, \exp\left[-(x-\eta)^2/2\sigma^2\right],$$

regardless of the individual $p(x_i)$'s.

1.8.2 Stochastic Continuity

Having established the definitions of convergence of stochastic sequences, the properties of limits necessary to define stochastic differentiation and integration must be explored. Stochastic continuity deals with the way that the limits are defined according to the ways they are reached.

In the *deterministic case*, continuity of $X(t)$ is uniquely defined if

$$\lim_{\epsilon\to 0\pm} X(t+\epsilon) = X(t), \qquad \forall\, X(t), \tag{1.40}$$

where if $\epsilon \to 0+$ it is upper semicontinuous, and if $\epsilon \to 0-$ it is lower semicontinuous.

Continuity* w.p. *1 is Restrictive. A stochastic process $X(t)$ is a family (ensemble) of functions. If each member is continuous at time t, the process is continuous at t for *almost all* ζ (experiments), or

$$\text{Prob}\left[\lim_{\tau\to 0} X(t+\tau) = X(t)\right] = 1. \tag{1.41}$$

Continuity in the m.s. A stochastic process is continuous in m.s. if

$$\lim_{\tau\to 0} E\left\{\| X(t+\tau) - X(t)\|^2\right\} = 0, \tag{1.42}$$

but $R_{xx}(0) = E\{X(t)^2\}$ and the process is continuous if R_{xx} is continuous.

Continuity in the Mean. A stochastic process is continuous in the mean if

$$\lim_{\tau\to 0} E\{X(t+\tau)\} = E\{X(t)\}. \tag{1.43}$$

With these definitions in mind, one may proceed to the next subject.

1.8.3 Stochastic Differentiation

The derivative of a function $X(t)$ is defined as a limit:

$$X'(t) = dX(t)/dt = \lim_{\tau\to 0} \frac{[X(t+\tau) - X(t)]}{\tau}. \tag{1.44}$$

If this limit exists for all experiments ζ of the process, then the process has a regular derivative. However, this may not be always true for all the ζ's. Then a derivative may be defined according to the preferred limit, for example, w.p. 1. In the mean, or as preferred hereafter, in the mean-square sense (m.s.s.), if

$$\lim_{\tau\to 0}\left[E\left\{\frac{X(t+\tau) - X(t)}{\tau} - X'(t)\right\}^2\right]. \tag{1.45}$$

Theorem 1.2 A stationary stochastic process $X(t)$ is differentiable in m.s. iff $R(\tau)$, the autocorrelation function, has derivatives of order up to 2.

Proof: The differentiability condition implies

$$dR(0)/d\tau = 0, \qquad \lim_{\tau\to 0}\frac{R(\tau) - R(0)}{\tau^2} = d^2R(0)/d\tau^2. \tag{1.46}$$

The condition to be proved is written as

$$\lim_{\tau \to 0} \left[E\left\{ \frac{X(t+\tau) - X(t)}{\tau} - X'(t) \right\}^2 \right]$$
$$= \lim_{\epsilon, \tau \to 0} \left[E\left\{ \frac{X(t+\tau) - X(t)}{\tau} - \frac{X(t+\epsilon) - X(t)}{\epsilon} \right\}^2 \right] = 0. \quad (1.47)$$

Then $\gamma = \epsilon, \tau$; it is easy to show that

$$\lim_{\gamma \to 0} \left[E\left\{ \frac{X(t+\gamma) - X(t)}{\gamma} \right\}^2 \right] = \lim_{\gamma \to 0} \frac{[R(0) - R(\gamma)]}{\gamma^2} = -\tfrac{1}{2} R''(0),$$

$$\lim_{\epsilon, \tau \to 0} \left[E\left\{ \frac{X(t+\tau) - X(t)}{\tau} \frac{X(t+\epsilon) - X(t)}{\epsilon} \right\}^2 \right] = -R''(0).$$

Substituting back to (1.47), one gets

$$\lim_{\tau \to 0} \left[E\left\{ \frac{X(t+\tau) - X(t)}{\tau} - X'(t) \right\}^2 \right] = -2R''(0) + 2R''(0) = 0. \quad (1.48)$$

Q.E.D.

A generalization of the theorem for a nonstationary process $X(t)$ to be differentiable in m.s. is that the second partial exists for $t_1 = t_2$:

$$\partial^2 R(t_1, t_2)/\partial t_1 \, \partial t_2. \quad (1.49)$$

Taylor-Series Expansion. If the autocorrelation function $R(\tau)$ of a stationary process exists, then $X(t)$ can be expanded in Taylor series; for example, if

$$R(\tau) = \sum_0^\infty R^{(n)}(0)(\tau^n/n!),$$

then

$$X(t+\tau) = \sum_0^\infty x^{(n)}(t)(\tau^n/n!). \quad (1.50)$$

Finally, the differential of a Wiener–Lévy process will be established by an example that demonstrates its dependence on second differentials.

Brownian-Motion Example. Consider the stochastic process $X(t)$, defined by

$$dX(t)/dt = W(t), \qquad X(0) \sim N(x_0, 0),$$
$$E\{W(t)W(\tau)\} = q\,\delta(t-\tau).$$

The variance of the process must satisfy the matrix Riccati equation, as explained in Chapter 3 on state estimation, for $f = 0$, $g = 1$:

$$dp/dt = 2fp + g^2 q \Rightarrow dp/dt = q, \qquad p(0) = 0.$$

Integrating,

$$p(t) = qt,$$

and

$$E\{X(t)\} = 0; \sigma^2 = E\{X(t)^2\} = p(t) = qt.$$

The variance of the process is proportional to t.

The *equivalent variations of* $X(t)$ are

$$\begin{aligned} \delta X &= \partial X/\partial t\,\delta t + \tfrac{1}{2}\partial^2 X/\partial t^2\,\delta t^2 + \partial X/\partial x\,\delta x + \tfrac{1}{2}\partial^2/\partial x^2\,\delta x^2 + \cdots \\ &= \partial X/\partial t\,\delta t + \partial X/\partial x\,\delta x + \tfrac{1}{2}\partial^2 X/\partial x^2\,\alpha\,\delta t + o(\delta t^2). \end{aligned} \tag{1.51}$$

As $\delta t \to dt$ and $\delta x \to dx$, the infinitesimal generator replacing the normal differential, with differentials of equal order, becomes

$$dX = \partial X/\partial t\,dt + \partial X/\partial x\,dx + \tfrac{1}{2}\partial^2 X/\partial x^2\,\alpha\,dt. \tag{1.52}$$

This establishes the difference between deterministic and stochastic differentiation.

1.8.4 Stochastic Integration

Given a stochastic process $X(\zeta, t)$, if the Riemann integral exists for all ζ,

$$S(\zeta) = \int_a^b X(\zeta, t)\,dt, \qquad \forall\, x(\zeta, t), \tag{1.53}$$

then the process is Riemann integrable and $S(\zeta)$ is a random variable. However, if it does *not* exist for every ζ, the integral S is defined by

$$\lim_{\Delta t_i \to 0} E\left\{\left[S - \sum_i X(t_i)\,\Delta t_i\right]^2\right\} = 0, \qquad \text{in the m.s.s.}, \tag{1.54}$$

as a *stochastic integral*. In order to evaluate the stochastic integral S, one should determine its mean value and variance, which involve linear operators, because it is almost hopeless to find the density $p(s)$:

$$E\{S\} = \int_a^b E\{X(t)\}\,dt = \int_a^b X(t)\,dt, \tag{1.55}$$

$$\begin{aligned} E\{S^2\} &= E\left\{\int_a^b X(t_1)\,dt_1 \int_a^b X(t_2)\,dt_2\right\} = E\left\{\int_a^b\int_a^b X(t_1)X(t_2)\,dt_1\,dt_2\right\} \\ &= E\{S^2\} = \int_a^b\int_a^b E\{X(t_1)X(t_2)\,dt_1\,dt_2\} = \int_a^b\int_a^b R(t_1,t_2)\,dt_1\,dt_2, \end{aligned} \tag{1.56}$$

and

$$\sigma_s^2 = \int_a^b\int_a^b C(t_1,t_2)\,dt_1\,dt_2. \tag{1.57}$$

In the case of *sample moments*, for an $X(t)$ that is a stationary process, the integral is formed:

$$S = (1/2T)\int_{-T}^{T} X(t)\,dt, \tag{1.58}$$

and if η is the mean value and $C(\tau)$ the autocovariance of $X(t)$,

$$\begin{aligned} E\{S\} &= \eta, \\ \sigma_s^2 &= (1/4T^2)\int_{-T}^{T}\int_{-T}^{T} C(t_1 - t_2)\,dt_1\,dt_2 \\ &= (1/2T)\int_{-2T}^{2T} (1 - |\tau|/2T)C(\tau)\,d\tau. \end{aligned}$$

1.9 ENTROPY

Entropy may be used as a measure of uncertainty. This concept was introduced by Shannon in the 1940s, as an extension of Boltzmann's idea, to measure the uncertainty of transmission of information in his celebrated *information theory*:

$$H = \int_\Omega p(s)\ln p(s)\,ds, \tag{1.59}$$

where $p(s)$ is a density function defined over the space Ω of the information signals transmitted. The space Ω can be partitioned into a collection of mutually exclusive events U, covering Ω, with an entropy measure $H(U)$, expressing the uncertainty of being in a particular partition U_i, $i = 1, \ldots, n$. Some properties of $H(U)$ are

1. $H(U)$ is a continuous function of $p(U_i) = p_i$.
2. If $p_1 \cdots p_N = 1/N$, then $H(U)$ is an increasing function of N.
3. If a new partition V is defined by further subdividing Ω into $N + M$ partitions, then $H(V) \geq H(U)$.

The concept of entropy and its applications to stochastic estimation, optimal control, et cetera, will be further explored in the following. A major emphasis will be given to this concept because it may represent a global measure of uncertainty unifying various unrelated disciplines.

1.10 CORRELATION FUNCTIONS

Correlation functions are very important second moments of a stochastic process, not only because they describe the dependence of the random values at different times, but because they provide a deterministic means to describe the performance of linear systems. They are defined as autocorrelation functions,

$$R_{XX}(t_1, t_2) = E\{X(t_1)X(t_2)\}, \tag{1.60}$$

when they relate two values of the same process at different times, and as cross-correlation functions,

$$R_{XY}(t_1, t_2) = E\{X(t_1)Y(t_2)\}, \tag{1.61}$$

when they relate two values of two different processes at different times.

1.10.1 Autocorrelation Functions

For a general stochastic process the autocorrelation function is defined as

$$\begin{aligned} R_{XX}(t_1, t_2) &= E\{X(t_1)X(t_2)\} \\ &= \int_{-\infty}^{\infty}\int_{-\infty}^{\infty} x(t_1)x(t_2)p(x_1, t_1; x_2, t_2)\,dx_1, dx_2. \end{aligned} \tag{1.62}$$

If the process is stationary the autocorrelation function reads

$$R_{XX}(t_1, t_1 + \tau) = E\{X(t_1)X(t_1 + \tau)\} = R_{XX}(\tau). \tag{1.63}$$

1.10.2 Time Autocorrelation Functions

They are defined as a time average over a sample function:

$$\Re_{XX}(\tau) = \langle X(t)X(t+\tau)\rangle = \lim_{T\to\infty} (1/2T)\int_{-T}^{T} X(t)X(t+\tau)\,d\tau. \quad (1.64)$$

For ergodic processes

$$R_{XX}(\tau) = \Re_{XX}(\tau). \quad (1.65)$$

1.10.3 Correlation Coefficient of Stochastic Processes

Let $X(t)$ be a stationary stochastic process with $E\{X(t)\} = 0$, and $Y(t)$,

$$\begin{aligned} Y(t) &= X(t) - \rho X(t+\tau), \\ E\{Y(t)\} &= 0, \\ E\{Y(t)^2\} &= E\{[X(t) - \rho X(t+\tau)]^2\} \\ &= E\{X(t)^2 - 2\rho X(t)X(t+\tau) + \rho^2 X(t+\tau)^2\} \\ &= \sigma_X^2 - 2\rho R_{XX}(\tau) + \rho^2\sigma_X^2 = \sigma_Y^2. \end{aligned}$$

Find ρ that minimizes σ_Y^2,

$$d\sigma_Y^2/d\rho = -2R_{XX}(\tau) + 2\rho\sigma_X^2 = 0,$$

which is defined as the correlation coefficient for the process $X(t)$;

$$\rho = R_{XX}(\tau)/\sigma_X^2. \quad (1.66)$$

Because

$$R_{XX}(0) = \sigma_X^2 = \max_{\tau} R_{XX}(T) \Rightarrow \max|\rho| = 1, \qquad -1 \le \rho \le 1 \quad (1.67)$$

or

$$\begin{aligned} \rho &= 1 \Rightarrow \text{complete correlation}, \\ \rho &= 0 \Rightarrow \text{complete uncorrelation}, \\ \rho &= -1 \Rightarrow \text{identical waveforms of opposite sign}. \end{aligned}$$

EXAMPLE Show that the stochastic process

$$X(t) = A\cos(\omega t + \theta); \qquad \text{with } A, \omega = \text{const.} \quad \text{and}$$
$$\mathrm{p}(\theta) = 1/2\pi, \theta \in [0, 2\pi],$$

is ergodic in the wide sense, and has an autocorrelation function

$$R_{XX}(\tau) = \tfrac{1}{2}A^2 \cos \omega T.$$

In order to show ergodicity first evaluate the ensemble and sample means:

$$E\{X(t)\} = AE\{\cos(\omega t + \theta)\} = 0,$$

$$\langle X(t)\rangle = A \lim_{T\to\infty} (1/2T)\int_{-T}^{T} \cos(\omega t + \theta)\, dt = 0.$$

Then evaluate the respective autocorrelation functions:

$$\begin{aligned} R_{XX}(T) &= E\{A\cos(\omega t + \theta) A\cos(\omega t + \omega\tau + \theta)\} \\ &= \tfrac{1}{2}A^2 E\{\cos(2\omega t + \omega\tau + 2\theta) + \cos(\omega\tau)\} = \tfrac{1}{2}A^2 \cos \omega\tau, \end{aligned}$$

$$\begin{aligned} \Re_{XX}(\tau) &= \lim_{T\to\infty} (1/2T)\int_{-T}^{T} A\cos(\omega t + \theta) A\cos(\omega t + \omega\tau + \theta)\, dt \\ &= \tfrac{1}{2}A^2 \lim_{T\to\infty} (1/2T)\left[\int_{-T}^{T} \cos(\omega\tau) + \int_{-T}^{T} \cos(\omega t + \omega\tau + \theta)\right] dt \\ &= \tfrac{1}{2}A^2 \cos \omega\tau. \end{aligned}$$

Therefore, the process is ergodic in the wide sense and has an autocorrelation function $\frac{1}{2}A^2 \cos \omega\tau$.

1.10.4 Properties of Stationary Autocorrelation Functions

1. $R_{XX}(0) = E\{X^2\}$.

Proof:

$$R_{XX}(\tau) = \int_{-\infty}^{\infty} x(t)x(t+\tau)p(x)\, dx \Rightarrow R_{XX}(0)$$

$$= \int_{-\infty}^{\infty} x(t)^2 p(x)\, dx = E\{X^2\}.$$

Q.E.D.

2. $R_{XX}(\tau) = R_{XX}(-\tau)$.

Proof:

$$\begin{aligned} R_{XX}(-\tau) &= \int_{-\infty}^{\infty} x(t)x(t-\tau)p(x)\, dx \\ &= \int_{-\infty}^{\infty} x(s+\tau)x(s)p(x)\, dx = R_{XX}(\tau). \end{aligned}$$

Q.E.D.

3. $|R_{XX}(\tau)| \le R_{XX}(0)$.

Proof:

$$E\{[X(t_1) \pm X(t_2)]^2\} = E\{X(t_1)^2 + X(t_2)^2 \pm 2X(t_1)X(t_2)\} \geq 0$$

$$\Rightarrow E\{X(t_1)^2 + X(t_2)^2\} = 2R_{XX}(0) \geq |E\{2X(t_1)X(t_2)\}|$$

$$= 2|R_{XX}(\tau)|. \qquad \text{Q.E.D.}$$

4. If $X(t)$ has a constant value component, $R_{XX}(\tau)$ also has a constant value component.

Proof: Let $X(t) = X + N(t)$ [$N(t)$ a zero-mean process],

$$R_{XX}(\tau) = E\{[X + N(t)][X + N(t + \tau)]\} = X^2 + R_{NN}(\tau). \qquad \text{Q.E.D.}$$

5. If $X(t)$ has a periodic component, so does $R_{XX}(\tau)$.

Proof: Let

$$X(t) = A\cos(\omega t + \theta) + Y(t), \qquad A, \omega = \text{const.},$$

$$p(\theta) = 1/2\pi, \quad 0 \leq \theta \leq 2\pi;$$

$$R_{XX}(\tau) = E\{A\cos(\omega t + \theta)A\cos(\omega t + \omega\tau + \theta)\} + E\{Y(t)Y(t + \tau)\}$$

$$= \tfrac{1}{2}A^2\cos\omega\tau + R_{YY}(\tau). \qquad \text{Q.E.D.}$$

6. If $X(t)$ is zero-mean ergodic with no periodic components,

$$\lim_{\tau \to \infty} R_{XX}(\tau) \to 0.$$

Proof:

$$|R_{XX}(\tau)| \leq R_{XX}(0) \Rightarrow \text{bounded}; \qquad R_{XX}(\tau) = \Re_{XX}(\tau) \to 0, \text{ as } \tau \to \infty.$$

Q.E.D.

Considerable information about a stochastic process $X(t)$ can be recovered by observing its autocorrelation function; for example, the mean and variance,

$$R_{XX}(\tau) = \overline{X}^2 + \tfrac{1}{2}A^2\cos\omega\tau + B^2\exp[-|\tau|]\cos\tau,$$

yields

$$E\{X(t)\} = \pm\overline{X}, \qquad E\{X^2\} = R_{XX}(0) = \overline{X}^2 + \tfrac{1}{2}A^2 + B^2,$$
$$\sigma_X^2 = E\{X^2\} - \overline{X}^2 = \tfrac{1}{2}A^2 + B^2.$$

1.10.5 Cross-Correlation Functions

For two general stochastic processes $X(t)$ and $Y(t)$, the *cross-correlation functions* $R_{XY}(t_1, t_2)$ and $R_{YX}(t_1, t_2)$ are defined, respectively, as;

$$R_{XY}(t_1, t_2) = E\{X(t_1)Y(t_2)\},$$
$$R_{YX}(t_1, t_2) = E\{Y(t_1)X(t_2)\}. \tag{1.68}$$

For stationary processes and $t_2 - t_1 = \tau$,

$$R_{XY}(t_1, t_2) = R_{XY}(\tau), \tag{1.69}$$
$$R_{YX}(t_1, t_2) = R_{YX}(\tau). \tag{1.70}$$

It is obvious that, for either case, $R_{XY}(t_1, t_2) \equiv R_{YX}(t_1, t_2)$.

The *time cross-correlation functions* are defined, respectively, as

$$\mathfrak{R}_{XY}(\tau) = \lim_{T\to\infty} (1/2T)\int_{-T}^{T} x(t)y(t+\tau)\,dt, \tag{1.71}$$

$$\mathfrak{R}_{YX}(\tau) = \lim_{T\to\infty} (1/2T)\int_{-T}^{T} y(t)x(t+\tau)\,dt. \tag{1.72}$$

Again, it is obvious that $\mathfrak{R}_{XY}(\tau) \neq \mathfrak{R}_{YX}(\tau)$.

For ergodic processes

$$R_{XY}(\tau) = \mathfrak{R}_{XY}(\tau), \tag{1.73}$$
$$R_{YX}(\tau) = \mathfrak{R}_{YX}(\tau). \tag{1.74}$$

1.10.6 Properties of Stationary Cross-Correlation Functions

1. $R_{XY}(0) = R_{YX}(0)$.

Proof: From the definition, $R_{XY}(0) = E\{X(t_1)Y(t_1)\} = R_{YX}(0)$. Q.E.D.

2. Cross-correlation functions are not even functions of τ, $R_{XY}(\tau) \neq R_{XY}(-\tau)$, but $R_{XY}(\tau) = R_{YX}(-\tau)$.

Proof:

$$R_{XY}(\tau) = E\{X(t)Y(t+\tau)\} = E\{X(u-\tau)Y(u)\} = R_{YX}(-\tau),$$

$$\text{for } u = t + \tau. \quad \text{Q.E.D.}$$

3. $|R_{XY}(\tau)| \leq \frac{1}{2}[R_{XX}(0) + R_{YY}(0)]$ and $\max R_{XY}(\tau) \neq R_{XY}(0)$.

Proof: Let $X(t_1) = X_1$, $X(t_2) = X_2$; $Y(t_1) = Y_1$, $Y(t_2) = Y_2$, and

$$0 \leq E\{(X_1 \pm Y_2)^2\} = E\{X_1^2\} + E\{Y_2^2\} \pm 2E\{X_1Y_2\} = R_{XX}(0) + R_{YY}(0)$$

$$\pm 2R_{XY}(\tau) \Rightarrow |R_{XY}(\tau)| \leq \tfrac{1}{2}[R_{XX}(0) + R_{YY}(0)].$$

The worst case yields

$$\max_{\tau} R_{XY}(\tau) = \tfrac{1}{2}[R_{XX}(0) + R_{YY}(0)] \neq R_{XY}(0). \qquad \text{Q.E.D.}$$

4. $[R_{XY}(\tau)]^2 \leq R_{XX}(0)R_{YY}(0)$.

Proof: Let

$$E\{[X(t) + \alpha Y(t+\tau)]^2\} = R_{XX}(0) + 2\alpha R_{XY}(\tau) + \alpha^2 R_{YY}(0) \geq 0.$$

The trinomial in α must not change sign; for example, its discriminant must be nonpositive:

$$R_{XY}(\tau)^2 - R_{XX}(0)R_{YY}(0) \leq 0$$

or

$$|R_{XY}(\tau)| \leq \sqrt{R_{XX}(0)R_{YY}(0)} \leq \tfrac{1}{2}[R_{XX}(0) + R_{YY}(0)]. \qquad \text{Q.E.D.}$$

5. If $X(t)$ and $Y(t)$ are stationary and statistically independent (i.i.d.),

$$R_{XY}(\tau) = E\{X(t)\}E\{Y(t)\} = R_{YX}(\tau).$$

6. If $X(t)$ is stationary and differentiable in m.s.,

$$R_{X'X'}(\tau) = E\{X'(t)X'(t+\tau)\} = d^2R_{XX}(\tau)/d\tau^2.$$

Proof:

$$
\begin{aligned}
R_{XX'}(\tau) &= E\{X(t)X'(t+\tau)\} \\
&= E\left\{\lim_{\epsilon\to 0}\left[\frac{X(t)X(t+\tau+\epsilon) - X(t)X(t+\tau)}{\epsilon}\right]\right\} \\
&= \lim_{\epsilon\to 0}\{[R_{XX}(\tau+\epsilon) - R_{XX}(\tau)]/\epsilon\} = dR_{XX}(\tau)/d\tau.
\end{aligned}
$$

Repeating the process once more yields

$$R_{X'x'}(\tau) = dR_{XX'}(\tau)/d\tau = d^2/R_{XX}(\tau)/d\tau^2. \qquad \text{Q.E.D.}$$

This result was observed in the case of stochastic differentiation, for example, Eq. (1.46).

EXAMPLES OF CORRELATION FUNCTIONS The autocorrelation resulting from the weighted sum (difference) $Z(t)$ of two independent stochastic processes $X(t), Y(t)$,

$$Z(t) = X(t) + \alpha Y(t), \tag{1.75}$$

is given in general:

$$
\begin{aligned}
R_{ZZ}(\tau) &= E\{[X(t) \neq \alpha Y(t)][X(t+\tau) \pm \alpha Y(t+\tau)]\} \\
&= R_{XX}(\tau) + \alpha^2 R_{YY}(\tau) \pm \alpha[R_{XY}(\tau) + R_{YX}(\tau)].
\end{aligned} \tag{1.76}
$$

Particularly, if

$$X(t) = A\cos(\omega t + \theta), \quad A, \omega = \text{const., and} \quad p(\theta) = 1/2\pi,$$

with

$$E\{X(t)\} = 0 \quad \text{and} \quad R_{XX}(\tau) = \tfrac{1}{2}A\cos\omega\tau,$$

and $Y(t)$ is a binary signal with randomly spaced switching T, and

$$R_{YY}(\tau) = B^2\exp\{-\alpha|\tau|\}.$$

If $X(t)$ and $Y(t)$ are statistically independent,

$$R_{XY}(\tau) = R_{YX}(\tau) = E\{X(t)\}E\{Y(t)\} = 0$$

and

$$R_{ZZ}(\tau) = R_{XX}(\tau) + R_{YY}(\tau) = 1/2A^2 \cos \omega\tau + B^2 e^{-\alpha|\tau|}.$$

1.10.7 Covariance Functions

Covariance functions for scalar stochastic processes are defined as follows.

Autocovariance

$$\begin{aligned} C_{XX}(t, t+\tau) &= E\{[X(t) - \overline{X}(t)][X(t+\tau) - \overline{X}(t+\tau)]\} \\ &= R_{XX}(t, t+\tau) - E\{X(t)X(t+\tau)\}. \end{aligned}$$

For wide-sense stationary processes

$$C_{XX}(\tau) = R_{XX}(\tau) - \overline{X}^2 \quad \text{and} \quad \sigma_X^2 = R_{XX}(0) + \overline{X}^2. \tag{1.77}$$

Cross-Covariance

$$\begin{aligned} C_{XY}(t, t+\tau) &= E\{[X(t) - \overline{X}(t)][Y(t+\tau) - \overline{Y}(t+\tau)]\} \\ &= R_{XY}(t, t+\tau) - \overline{X}(t)\overline{Y}(t+\tau), \end{aligned}$$

and for wide-sense stationary processes

$$C_{XY}(\tau) = R_{XY}(\tau) - \overline{XY}, \tag{1.78}$$

similarly for $C_{YX}(t, t+\tau)$ and $C_{YX}(\tau)$. If the processes are uncorrelated,

$$C_{YX}(t, t+\tau) = 0 \Rightarrow R_{YX}(\tau) = \overline{XY}. \tag{1.79}$$

In the case of vector stochastic processes $X(t), Y(t)$, the correlation functions are defined, respectively, as matrices:

$$\begin{aligned} R_{XX}(\tau) &= E\{X(t)X(t+\tau)^T\}, \\ R_{XY}(\tau) &= E\{X(t)Y(t+\tau)^T\}, \\ C_{XX}(\tau) &= E\{[X(t) - \overline{X}][X(t+\tau) - \overline{X}]^T\}, \qquad \text{et cetera.} \end{aligned} \tag{1.80}$$

1.10.8 Discrete-Time Moments and Correlation Functions

Time or sample moments and correlation functions have been encountered previously for continuous stochastic processes, respectively,

$$\langle X^n \rangle = \lim_{T\to\infty} (1/2T) \int_{-T}^{T} X(t)^n \, dt,$$

$$\mathfrak{R}_{XX}(\tau) = \langle X(t) X(t+\tau) \rangle = \lim_{T\to\infty} (1/2T) \int_{-T}^{T} X(t) X(t+\tau) \, dt,$$

$$\mathfrak{R}_{XY}(\tau) = \lim_{T\to\infty} (1/2T) \int_{-T}^{T} X(t) Y(t+\tau) \, dt,$$

$$\mathfrak{R}_{YX}(\tau) = \lim_{T\to\infty} (1/2T) \int_{-T}^{T} Y(t) X(t+\tau) \, dt.$$

Equivalently, time moments and correlation functions may be defined for discrete stochastic processes respectively,

$$\langle X^n \rangle = \lim_{N\to\infty} (1/N) \sum_{i=1}^{N} X_i^n, \tag{1.81}$$

$$\mathfrak{R}_{XX}(k) = \langle X_n X_{n+k} \rangle = \lim_{N\to\infty} (1/N) \sum_{n=1}^{N} X_n X_{n+k}, \tag{1.82}$$

$$\mathfrak{R}_{XY}(k) = \lim_{N\to\infty} (1/N) \sum_{n=1}^{N} X_n Y_{n+k}, \tag{1.83}$$

$$\mathfrak{R}_{YX}(k) = \lim_{N\to\infty} (1/N) \sum_{n=1}^{N} Y_n X_{n+k}. \tag{1.84}$$

If the first and second time moments are equal to their ensemble equivalent ones, the process is also ergodic in the wide sense:

$$\langle X \rangle = \lim_{N\to\infty} (1/N) \sum_{i=1}^{N} X_i = \lim_{N\to\infty} (1/N) \sum_{i=1}^{N} X_i p(X_i) = E\{X\},$$

$$\langle X^2 \rangle = \lim_{N\to\infty} (1/N) \sum_{i=1}^{N} X_i^2 = \lim_{N\to\infty} (1/N) \sum_{i=1}^{N} X_i^2 p(X_i) = E\{X^2\}. \tag{1.85}$$

1.11 LINEAR SYSTEMS WITH STOCHASTIC INPUTS

Linear systems are defined by their input-output relation:

$$y(t) = L[x(t)], \tag{1.86}$$

where L is the linear operator satisfying the linearity conditions, for a particular experiment ζ;

$$L[a_1(\zeta)x_1(t) + a_2(\zeta)x_2(t)] = a_1(\zeta)L[x_1(t)] + a_2(\zeta)L[x_2(t)],$$

and

$$L[a(\zeta)x(t)] = a(\zeta)L[x(t)]. \tag{1.87}$$

Theorem 1.3 For the stochastic processes $Y(t)$, $X(t)$ related by (1.86),

$$E\{Y(t)\} = L[E\{X(t)\}]. \tag{1.88}$$

Proof: Expectation is a linear operator (integral), and linear operators commute. Q.E.D.

1.11.1 Correlation Functions

Theorem 1.4 For the stochastic processes $Y(t)$, $X(t)$, related by (1.86),

$$R_{YX}(t_1, t_2) = L_{t1}[R_{XX}(t_1, t_2)], \tag{1.89}$$

where L_{t1} is the linear operator associated with time t_1.

Proof:

$$\begin{aligned} R_{YX}(t_1, t_2) &= E\{Y(t_1)X(t_2)\} \\ &= E\{L_{t1}[X(t_1)]X(t_2)\} = L_{t1}[E\{X(t_1)X(t_2)\}] \\ &= L_{t1}[R_{XX}(t_1, t_2)]. \end{aligned}$$

Q.E.D.

Corollary 1.1 Similarly, L_{t2} is the linear operator associated with time t_2;

$$R_{XY}(t_1, t_2) = L_{t2}[R_{XX}(t_1, t_2)]. \tag{1.90}$$

For a stationary process $X(t)$, passing through a linear system with impulse response $h(t)$, with $h(t) = 0$ for $t < 0$ due to causality, and generating an output $Y(t)$, the convolution integral applies:

$$Y(t) = \int_0^\infty h(\sigma)X(t - \sigma)\, d\sigma, \tag{1.91}$$

and the mean value is

$$E\{Y(t)\} = \int_0^\infty h(\sigma)E\{X(t-\sigma)\}\,d\sigma = E\{X\}\int_0^\infty h(\sigma)\,d\sigma. \quad (1.92)$$

The mean value of the output does not always provide meaningful information, because $E\{X\}$ may be zero. The cross-correlation functions, however, give a linear relation between input and output:

$$\begin{aligned} R_{XY}(\tau) &= E\{X(t_1)Y(t_2)\} = E\left\{X(t_1)\int_0^\infty h(\sigma)X(t_1+\tau-\sigma)\,d\sigma\right\} \\ &= \int_0^\infty h(\sigma)E\{X(t_1)X(t_1+\tau-\sigma)\}\,d\sigma \\ &= \int_0^\infty h(\sigma)R_{XX}(\tau-\sigma)\,d\sigma = R_{XX}(\tau)*h(\tau), \end{aligned} \quad (1.93)$$

where the asterisk is the symbol of convolution, and $\tau = t_2 - t_1$. Similarly,

$$\begin{aligned} R_{YX}(\tau) &= E\{Y(t_1)X(t_2)\} = E\left\{\int_0^\infty h(\sigma)X(t_1-\sigma)\,d\sigma\,X(t_1+\tau)\right\} \\ &= \int_0^\infty h(\sigma)E\{X(t_1-\sigma)X(t_1+\tau)\}\,d\sigma \\ &= \int_0^\infty h(\sigma)R_{XX}(\tau+\sigma)\,d\sigma = R_{XX}(\tau)*h(-\tau), \end{aligned} \quad (1.94)$$

where the property $R_{XX}(\tau) = R_{XX}(-\tau)$ was used.

Theorem 1.5 For the stationary stochastic processes $Y(t)$, $X(t)$, representing the output and input of the linear system with $h(t)$, its impulse response

$$R_{YY}(\tau) = \int_0^\infty R_{XY}(\tau+\alpha)h(\alpha)\,d\alpha = R_{XX}(\tau)*h(-\tau)*h(\tau). \quad (1.95)$$

Proof:

$$\begin{aligned} R_{YY}(\tau) &= E\{Y(t)Y(t+\tau)\} \\ &= E\left\{\int_0^\infty h(\alpha)X(t-\alpha)Y(t+\tau)\,d\alpha = \int_0^\infty R_{XY}(\tau+\alpha)h(\alpha)\,d\alpha\right\} \\ &= R_{XX}(\tau)*h(-\tau)*h(\tau). \end{aligned}$$

Q.E.D.

1.11.2 Sums of Signals

If the input process $X(t)$, is composed of the sum of several signals,

$$X(t) = \sum_{n=1}^{N} a_n X_n(t), \tag{1.96}$$

then, due to the linearity of the operators involved, the output $Y(t)$ is

$$Y(t) = L[X(t)] = L\left[\sum_{n=1}^{N} a_n X_n(t)\right] = \sum_{n=1}^{N} a_n L[X_n(t)]$$

$$= \sum_{n=1}^{N} a_n Y_n(t), \tag{1.97}$$

when $Y_n(t) = L[X_n(t)]$.

EXAMPLE Given the RC network of Figure 1.9, and a stationary white-noise process $X(t)$ with X as mean and S_0 as variance, find the mean and autocorrelation function of the output $Y(t)$.

If the process $X(t)$ is white noise, as discussed later, its mean is X and autocorrelation function $R_{XX}(\tau) = S_0\delta(\tau)$, where $\delta(\tau)$ is the Dirac delta function. Then the mean value of the output is

$$Y = E\left\{\int_0^{\infty} h(\sigma) X(t-\sigma)\, d\sigma\right\} = \int_0^{\infty} h(\sigma) E\{X(t-\sigma)\}\, d\sigma$$

$$= (\overline{X}/RC)\int_0^{\infty} \exp\{-\sigma/RC\}\, d\sigma,$$

$$\overline{Y} = (\overline{X}/RC)[-RC]\exp\{-\tau/RC\}\big|_0^{\infty} = \overline{X}.$$

The autocorrelation function of the output $R_{YY}(\tau)$ is computed from the

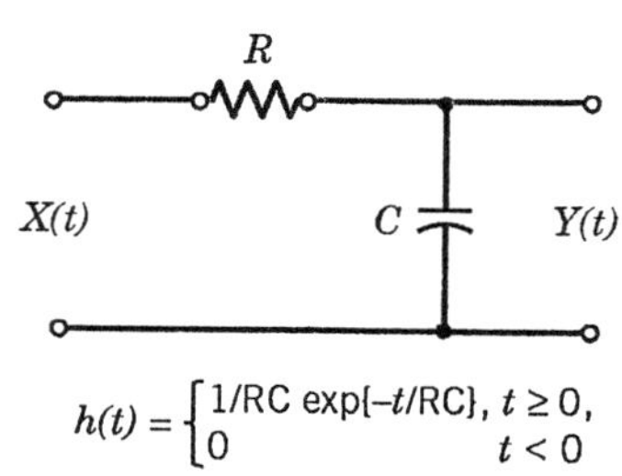

Fig. 1.9 Low Pass filter

cross correlation $R_{XY}(\tau)$ of the previous theorem;

$$\begin{aligned} R_{XY}(\tau) &= \int_0^\infty h(\sigma) R_{XX}(\tau - \sigma)\, d\sigma = S_0 \int_0^\infty h(\sigma) \delta(\tau - \sigma)\, d\sigma \\ &= S_0/RC \exp\{-\tau/RC\}, \end{aligned}$$

and the autocorrelation function of the output is, for $\tau \geq 0$,

$$\begin{aligned} R_{YY}(\tau) &= \int_0^\infty h(\sigma) R_{XY}(\tau + \sigma)\, d\sigma \\ &= S_0/(RC)^2 \int_0^\infty \exp\{-\sigma/RC\} \exp\{-\tau - \sigma/RC\}\, d\sigma \\ &= S_0/(RC)^2 \exp\{-\tau/RC\} \int_0^\infty \exp\{-2\sigma/RC\}\, d\sigma \\ &= S_0/2(RC) \exp\{-\tau/RC\}. \end{aligned}$$

The same result would be obtained if the double convolution formula of the theorem was applied:

$$\begin{aligned} R_{YY}(\tau) &= R_{XX}(\tau) * h(-\tau) * h(\tau) \\ &= \int_0^\infty \int_0^\infty h(\sigma) h(\alpha) R_{XX}(\alpha - \tau - \sigma)\, d\alpha\, d\sigma \\ &= \int_0^\infty \int_0^\infty h(\sigma) h(\alpha) S_0\, \delta(\alpha - \tau - \sigma)\, d\alpha\, d\sigma \\ &= S_0/(RC)^2 \int_0^\infty \exp\{-\sigma/RC\} \exp\{-\tau - \sigma/RC\}\, d\sigma \\ &= S_0/2(RC) \exp\{-\tau/RC\}. \end{aligned}$$

For $\tau < 0$ the autocorrelation function is

$$\begin{aligned} R_{YY}(\tau) &= S_0/(RC)^2 \int_{-\tau}^\infty \exp\{-\sigma/RC\} \exp\{-\tau - \sigma/RC\}\, d\sigma \\ &= (S_0/2RC) \exp\{\tau/RC\}, \end{aligned}$$

and the autocorrelation function of the output is given by

$$R_{YY}(\tau) = (S_0/2RC) \exp\{-|\tau|/RC\}, \qquad -\infty < \tau < \infty.$$

The variance of the output is given by

$$\sigma_Y^2 = R_{YY}(0) - \bar{Y}^2 = (S_0/2RC) - \bar{X}^2.$$

1.12 WHITE NOISE

A stochastic process $X(t)$ with an autocorrelation function $R_{XX}(\tau) = S_0\,\delta(\tau)$ is called white noise. The process is stationary so that its density function does not depend on time, and is uncorrelated at each time instant. An independent and identically distributed process is white noise, and the converse is true, if the process is also Gaussian.

EXAMPLE The output of an integrator, $h(t) = u(t) =$ step function, excited with white noise, is given:

$$Y(t) = \int_0^t X(\sigma)\,d\sigma.$$

The cross-correlation function of the output is

$$\begin{aligned} R_{YX}(t_1, t_2) &= E\{Y(t_1)X(t_2)\} = \int_0^{t_1} E\{X(\sigma)X(t_2)\}\,d\sigma \\ &= \int_0^{t_1} R_{XX}(\sigma, t_2)\,d\sigma \\ &= \int_0^{t_1} \delta(\sigma - t_2)\,d\sigma = u(t_1 - t_2), \qquad t_1, t_2 \geq 0, \end{aligned}$$

and

$$R_{YY}(t_1, t_2) = \begin{cases} t_2, & 0 \leq t_2 \leq t_1, \\ t_1, & 0 \leq t_1 \leq t_2. \end{cases}$$

If $X(t)$ is *thermal noise*, then $W(t)$ is a *Wiener–Lévy* process with

$$E\{X(t)\} = 0, \qquad E\{X(t)^2\} = ts^2/T = \alpha t$$

and

$$W(t) = \lim_{T \to 0} X(t), \Rightarrow E\{W(t)\} = 0,\ E\{W(t)^2\} = \alpha t,$$

$$R_{WW}(t_1, t_2) = \begin{cases} \alpha t_2 t_1 \geq t_2, \\ \alpha t_1 t_2 \geq t_1. \end{cases}$$

Therefore white Gaussian noise is the density of a Wiener–Lévy process.

1.13 SPECTRAL DENSITIES

Spectral densities are the Fourier (Laplace) transforms of correlation functions. They find considerable applications in signal analysis and processing. Here are some basic properties of transform theory. Let $x(t)$ be a sample function of a stochastic process.

Fourier Transform Pair.

$$X(\omega) = \int_{-\infty}^{\infty} x(t) e^{-j\omega t}\, dt,$$

$$x(t) = (1/2\pi) \int_{-\infty}^{\infty} X(\omega) e^{j\omega t}\, dt. \tag{1.98}$$

Parceval's Theorem.

$$\int_{-\infty}^{\infty} x(t) y(t)\, dt = (1/2\pi) \int_{-\infty}^{\infty} X(\omega) Y(-\omega)\, d\omega$$

or

$$\int_{-\infty}^{\infty} x(t)^2\, dt = (1/2\pi) \int_{-\infty}^{\infty} |X(\omega)|^2\, d\omega. \tag{1.99}$$

Truncated Functions. Truncated functions are created to avoid existence problems as $t \to \infty$.

$$f_T(t) = \begin{cases} f(t), & |t| \le T < \infty, \\ 0, & |t| > T. \end{cases} \tag{1.100}$$

Parceval's theorem applied on a truncated function yields

$$(1/2T) \int_{-T}^{T} f_T(t)^2\, dt = (1/4\pi T) \int_{-\infty}^{\infty} |F(\omega)|^2\, d\omega. \tag{1.101}$$

1.13.1 Moments of Truncated Functions

Using Parceval's theorem, the second moment of $x(t)$ is

$$E\left\{(1/2T) \int_{-T}^{T} x_T(t)^2\, dt\right\} = E\left\{(1/4\pi T) \int_{-\infty}^{\infty} |X(\omega)|^2\, d\omega\right\}$$

$$= (1/2T) \int_{-T}^{T} E\{x_T(t)^2\}\, dt,$$

and taking the limit as $T \to \infty$,

$$E\{X(t)^2\} = \lim_{T\to\infty} \left[(1/2T)\int_{-T}^{T} E\{x_T(t)^2\}\, dt\right]$$

$$= \lim_{T\to\infty} \left[(1/4\pi T)\int_{-\infty}^{\infty} E\{|X(\omega)|^2\}\, d\omega\right],$$

$$E\{X(t)^2\} = \left[(1/2\pi)\int_{-\infty}^{\infty} \lim_{T\to\infty} \frac{E\{|X(\omega)|^2\}}{2T}\, d\omega\right] = (1/2\pi)\int_{-\infty}^{\infty} S_{XX}(\omega)\, d\omega, \tag{1.102}$$

where

$$S_{XX}(\omega) = \lim_{T\to\infty} \frac{E\{|X(\omega)|^2\}}{2T}. \tag{1.103}$$

1.13.2 Transforms of Correlations

$$S_{XX}(\omega) = \int_{-\infty}^{\infty} R_{XX}(\tau) e^{-j\omega\tau}\, d\tau,$$

$$R_{XX}(\tau) = (1/2\pi)\int_{-\infty}^{\infty} S_{XX}(\omega) e^{j\omega\tau}\, d\omega, \tag{1.104}$$

and the second moment is given by

$$E\{X(t)^2\} = R_{XX}(0) = (1/2\pi)\int_{-\infty}^{\infty} S_{XX}(\omega)\, d\omega. \tag{1.105}$$

1.13.3 Cross-Spectral-Density Pairs

Cross-spectral-density pairs are defined by

$$S_{XY}(\omega) = \int_{-\infty}^{\infty} R_{XY}(\tau) e^{-j\omega\tau}\, d\tau,$$

$$R_{XY}(\tau) = (1/2\pi)\int_{-\infty}^{\infty} S_{XY}(\omega) e^{j\omega\tau}\, d\omega, \tag{1.106}$$

and

$$S_{YX}(\omega) = \int_{-\infty}^{\infty} R_{YX}(\tau) e^{-j\omega\tau}\, d\tau,$$

$$R_{YX}(\tau) = (1/2\pi) \int_{-\infty}^{\infty} S_{YX}(\omega) e^{j\omega\tau}\, d\omega. \qquad (1.107)$$

They may also be obtained from Parceval's theorem:

$$S_{XY}(\omega) = \lim_{T\to\infty} \frac{E\{X(\omega)Y(-\omega)\}}{2T}, \qquad (1.108)$$

$$S_{YX}(\omega) = \lim_{T\to\infty} \frac{E\{X(-\omega)Y(\omega)\}}{2T}. \qquad (1.109)$$

These functions represent complex values, for example, $X(\omega) = X_r(\omega) + jX_i(\omega)$.

1.13.4 Properties of Spectral Densities

1. $S_{XY}(\omega) = S_{YX}^*(\omega) =$ complex conjugate.
2. $\mathrm{Re}[S_{XY}(\omega)]$ is an even function of ω.
3. $\mathrm{Im}[S_{XY}(\omega)]$ is an odd function of ω.

1.13.5 Linear Systems

Based on the property that time convolution corresponds to frequency-domain multiplication,

$$y(t) = \int_0^{\infty} h(\sigma) x(t-\sigma)\, d\sigma \Rightarrow Y(\omega) = H(\omega) X(\omega),$$

the following relations are obtained from the above definitions:

$$S_{XY}(\omega) = \lim_{T\to\infty} \frac{E\{X(\omega)Y(-\omega)\}}{2T} = \lim_{T\to\infty} \frac{E\{X(\omega)H(-\omega)X(-\omega)\}}{2T}$$

$$= H(-\omega) \lim_{T\to\infty} \frac{E\{X(\omega)X(-\omega)\}}{2T} = H(-\omega) S_{XX}(\omega). \qquad (1.110)$$

The spectral density of the output is

$$S_{YY}(\omega) = \lim_{T\to\infty} \frac{E\{Y(\omega)Y(-\omega)\}}{2T} = \lim_{T\to\infty} \frac{E\{X(\omega)X(-\omega)\}}{2T} H(\omega)H(-\omega),$$

$$S_{YY}(\omega) = S_{XX}(\omega)|H(\omega)|^2. \tag{1.111}$$

1.13.6 Laplace Transform Expressions

Similar expressions may be obtained using the two-sided Laplace-transform pairs:

$$X(s) = \int_{-\infty}^{\infty} x(t)e^{-st}\,dt,$$

$$x(t) = (1/2\pi j)\int_{\alpha-j\infty}^{\alpha+j\infty} X(s)e^{st}\,dt. \tag{1.112}$$

The relations for linear systems are

$$S_{XY}(s) = H(s)S_{XX}(s),$$
$$S_{YX}(s) = H(-s)S_{XX}(s),$$
$$S_{YY}(s) = H(s)H(-s)S_{XX}(s). \tag{1.113}$$

Note. White noise has a spectral density $S_{XX}(s) = S_0$, an unrealistic value.

EXAMPLE Consider the low-pass filter previously treated, with impulse response

$$h(t) = 1/RC\exp\{-t/RC\}, \quad t \geq 0, \quad h(t) = 0, \quad t < 0.$$

Its transfer function is

$$H(s) = 1/[1 + RCs],$$

The mean value after convolution with the white noise input is

$$E\{Y(s)\} = H(s)E\{X(s)\},$$

input-output cross-spectral density is

$$S_{XY}(s) = H(s)S_{XX}(s) = S_0/[1 + RCs],$$

and the spectral density of the output is

$$S_{YY}(s) = H(s)H(-s)S_{XX}(s) = S_0/[1 + R^2C^2s^2].$$

With this the review of stochastic processes is completed.

PROBLEMS

1.1 The probability density function $p_X(x)$ of the random process $X(t)$ is normal with zero mean and variance σ^2. What is the probability density function $p_Y(y)$ of the random process given by $Y = 1/X$?

1.2 The probability density function $p_X(x)$ of the random process $X(t)$ is χ^2 with mean σ and variance 2σ. What is the probability density function $p_Y(y)$ of the random process given by $Y = X^2$?

1.3 Is the random binary reception process (a) continuous, (b) deterministic, (c) stationary, or (d) ergodic?

1.4 Is the Poisson process (a) continuous, (b) deterministic, (c) stationary, or (d) ergodic.

1.5 The output $Y(t)$ of a linear circuit is given by the convolution integral

$$Y(t) = \int_0^T h(\tau)X(t-\tau)\,d\tau,$$

where $h(t) = \exp[-t]$ and the input is a stochastic process

$$X(t) = A\cos(\omega t + \theta),$$

with A, ω = constants and θ a random variable uniformly distributed over $[0, \pi]$. Find the mean value Y and variance σ_Y^2 of the output process.

1.6

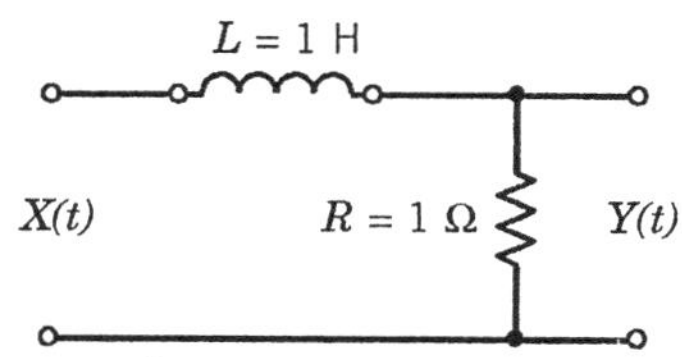

The stochastic process $X(t) = A\cos(t + \theta)$, with θ the only random variable with $p_\theta(\theta) = 1/2\pi$, $0 \le \theta \le 2\pi$, is applied to the above circuit. Find the mean value Y, and the variance σ_Y^2, of the output $Y(t)$.

1.7 A stationary process $X(t)$ has an autocorrelation function

$$R_{XX}(\tau) = 4\exp[-|\tau|]\cos(\pi\tau) + \cos(3\pi\tau) + 4.$$

Find the second moment of the process X^2 and its variance σ_X^2.

1.8 The process

$$y(t) = (1/2T)\int_{t-T}^{t+T} x(\alpha)\,d\alpha$$

represents the moving average of a process $x(t)$ in the interval $(t - T, t + T)$, and may be considered as the output of a system with impulse response a rectangular pulse on $[-T + T]$ with area $= 1$. If the autocorrelation function of the input is $R_{xx}(\tau)$ find the autocorrelation function of the output $R_{yy}(\tau)$.

1.9 The stochastic process

$$x(t) = A\cos(\omega_1 t + \phi_1) + B\cos(\omega_2 t + \phi_2),$$

where ϕ_1 and ϕ_2 are independent and identically uniformly distributed random variables over $[0, 2\pi]$, while all other parameters are constant. Find the autocorrelation function $R_{xx}(t_1, t_2)$ and the spectral density $S_{xx}(\omega)$.

REFERENCES

Cooper, G. R., and McGillem, C. D., *Probabilistic Methods of Signal and System Analysis*, Holt, Rinehart and Winston, New York, 1971.

Papoulis, A., *Probability, Random Variables, and Stochastic Processes* (3rd ed.), McGraw-Hill, New York, 1991.

Parzen, E., *Stochastic Processes*, Holden-Day, Inc., San Francisco, 1962.

2

PARAMETER ESTIMATION

2.1 POINT ESTIMATION

Point estimation provides the most *popular* value of a parameter X, of a sample function of a process with uncertainties, from a sequence of continuous- or discrete-time measurements Z. It is performed through

1. Sample statistics approach, for example, least squares (deterministic).
2. Probability theoretic methods (alternative, stochastic).

2.1.1 Sample Information

A. Continuous Time. Parameter estimation may be considered as the extraction of the average value of a sample function of measurements containing the parameter x:

$$z(t) = x + v(t).$$

Statistical Approach. The estimate of x is given by

$$\hat{x}(t) = (1/2T)\int_{-T}^{T} z(t)\, dt, \tag{2.1}$$

and the estimate of its variance σ_x^2 is

$$\hat{\sigma}_x^2 = (1/2T)\int_{-T}^{T} z(t)^2\, dt - x^2.$$

The process $v(t)$ is considered to be an unpredictable deterministic error. The measurement functions are considered as time functions and averaged through the estimation procedure.

Probabilistic Approach. The process $v(t)$ is considered to be a *zero-mean ergodic stochastic process* measured by a density function $p(n)$. Expressions (2.1) hold in both cases. However, the behavior of the estimator may be analyzed better by the probabilistic approach, using the analysis of stochastic processes. Such analysis is necessary, because expressions (2.1) are only approximations of the respective sample moments, in finite time:

$$\hat{x}(t) = (1/2T)\int_{-T}^{T} z(t)\,dt,$$

$$\hat{\sigma}_x^2 = (1/2T)\int_{-T}^{T} z(t)^2\,dt - x^2.$$

(a) Consider the expected value of the estimate,

$$\begin{aligned} E\{\hat{x}\} &= (1/2T)E\left\{\int_{-T}^{T} z(t)\,dt\right\} \\ &= (1/2T)\int_{-T}^{T} E\{z(t)\}\,dt = (x/2T)\int_{-T}^{T} dt = x. \end{aligned}$$

This property is called *unbiasedness*. It is easily shown that

$$\lim_{T\to\infty} E\{\hat{x}\} = \lim_{T\to\infty} (1/2T)\int_{-T}^{T} z(t)\,dt = \langle z(t)\rangle = x, \qquad \text{for ergodicity.}$$

(b) Consider the variance of the estimate σ_x^2:

$$\begin{aligned} \sigma_{\hat{x}}^2 = E\{\hat{x}^2\} - x^2 &= E\left\{(1/2T)\int_{-T}^{T} z(t)\,dt^2\right\} - (1/2T)\int_{-T}^{T} x\,dt^2 \\ &= (1/4T^2)\int_{-T}^{T}\int_{-T}^{T} E\{z(t)z(\alpha) - x^2\}\,dt\,d\alpha - x^2 \\ &= (1/4T^2)\int_{-T}^{T}\int_{-T}^{T} \left[R_{zz}(t-\alpha) - x^2\right]dt\,d\alpha, \end{aligned}$$

for an ergodic process. However, for $\tau = t - \alpha$,

$$\sigma_{\hat{x}}^2 = (1/4T^2)\int_{-T}^{T}\int_{-T-\tau}^{T-\tau}\left[R_{zz}(\tau) - x^2\right] dt\, d\sigma$$

$$= (1/4T^2)\int_{-T}^{T} dt\left[\int_{-T-\tau}^{0}\left[R_{zz}(\tau) - x^2\right] d\sigma + \int_{0}^{T-\tau}\left[R_{zz}(\tau) - x^2\right] d\sigma\right]$$

$$= (1/4T^2)\int_{-2T}^{2T} d\tau\big[[T+\tau]\left[R_{zz}(\tau) - x^2\right]$$

$$+[T-\tau]\left[R_{zz}(\tau) - x^2\right]\big] = (1/2T)\int_{-2T}^{2T}[T - |\tau|]\left[R_{zz}(\tau) - x^2\right] d\tau,$$

and the variance of the estimate of x, is written

$$\sigma_{\hat{x}}^2 = (1/2T)\int_{-2T}^{2T}[1 - |\tau|/T]\left[R_{zz}(\tau) - x^2\right] d\tau.$$

In the case that only the estimate of x is available, the *variance of the estimate* is written

$$\sigma_{\hat{x}}^2 = (1/2T)\int_{-2T}^{2T}[1 - |\tau|/T]\left[R_{zz}(\tau) - \left[(1/2T)\int_{-2T}^{2T} z(t)\, dt\right]^2\right] d\tau.$$

The estimate of the time-autocorrelation function is

$$\mathscr{R}_{zz}(\tau) = (1/2T)\int_{-T}^{T} z(t)z(t+\tau)\, dt$$

B. Discrete Time. Similar results are reached if the measurements of the process are obtained at discrete-time intervals:

$$z_i = x + v_i. \tag{2.2}$$

Statistical Approach. The estimate of x is given by

$$\text{sample mean} \rightarrow \hat{x} = (1/N)\sum_i^N z_i,$$

and the estimate of its variance σ_x^2 is

$$\text{sample variance} \rightarrow \hat{\sigma}_x^2 = (1/N)\sum_i^N (z_i - x)^2.$$

The process $\{v_i\}$ is considered to be an unpredictable deterministic error. The measurement values are considered as sequences of time and averaged through the estimation procedure.

Probabilistic Approach. The process $\{v_i\}$ is considered to be a *zero-mean ergodic stochastic process* measured by a probability function $p(v_i)$. Expressions (2.2) hold in both cases, as before. However, the behavior of the estimator may be analyzed better by the probabilistic approach, using the analysis of stochastic processes. Such analysis is necessary, because expressions (2.2) are only approximations of the respective sample moments, in finite time intervals N.

Assume that the error between measurement z_i and true or mean value x is v_i, a random variable with zero mean $E\{v_i\} = 0$ and finite variance $E\{v_i^2\} < \infty$. The *estimates* of the mean and the variance are again

$$\text{mean} \rightarrow \hat{x} = (1/N) \sum_i^N z_i,$$

$$\text{variance} \rightarrow \hat{\sigma}_x^2 = (1/N) \sum_i^N (z_i - x)^2.$$

The mean value and variance are unbiased:

$$E\{\hat{x}\} = (1/N) \sum_i^N E\{z_i\} = x,$$

$$E\{\hat{\sigma}_x^2\} = (1/N) \sum_i^N E(z_i - x)^2 = (1/N) \sum_i^N E\{\sigma_i^2\} = (1/N) \sum_i^N \operatorname{var}(z_i) = v_z^2.$$

The estimate of the time-autocorrelation function is

$$\mathscr{R}_{zz}(k) = 1/N \sum_i^N z_i z_{i+k}.$$

If the mean value x is not available and its estimate is used, the variance is biased. Assuming that $E\{v_k v_j\} = \delta_{kj}$ = Kroenecker delta:

$$\begin{aligned}
E\{\hat{\sigma}_x^2\} &= (1/N) \sum_i^N E\{(z_i - x)^2\} = (1/N) \sum_i^N E\left\{\left(z_i - (1/N) \sum_j^N z_j\right)^2\right\} \\
&= (1/N) \sum_i^N E\left\{\left(x + v_i - x - (1/N) \sum_j^N v_j\right)^2\right\} \\
&= (1/N) \sum_i^N E\left\{v_i^2 - 2/N v_i \sum_j^N v_j + (1/N^2) \sum_j^N \sum_k^N v_j v_k\right\} \\
&= (1/N) \sum_i^N \{v_z^2 - (1/N) v_z^2\} = (1/N) v_z^2 N(N-1)/N \\
&\rightarrow E\{\hat{\sigma}_x^2\} = v_z^2 (N-1)/N,
\end{aligned}$$

implying that the estimate is biased. The unbiased estimate of the variance is obtained by

$$\hat{v}_z^2 = N/(N-1)\hat{\sigma}_x^2 = (1/(N-1))\sum_i^N E\{(z_i - x)^2\}. \tag{2.3}$$

EXAMPLE

Given the sample-data nonlinear system see Figure 2.1, with input $X(t) \sim N(0, \sigma_x^2)$ and $Y(t) = X(t)^2$, find the number of samples N required to make the standard deviation $\sigma_{\hat{y}}$ of the estimate of the output less or equal to $0.1\overline{Y}$, the mean value of the output. Let

$$\overline{Y} = E\{Y\} = E\{X(t)^2\} = \sigma_x^2, \qquad E\{Y^2\} = E\{X(t)^4\} = 3\sigma_x^4,$$

$$\sigma_Y^2 = E\{Y^2\} - E\{Y\}^2 = 2\sigma_x^4.$$

After the discretization by Δt, the number N required to make $\sigma_y \leq 0.1E\{Y\}$ is calculated as follows:

$$\hat{Y} = (1/N)\sum_{i=1}^N Y_i \Rightarrow \sigma_{\hat{y}}^2 = E\{(\hat{Y} - E\{\hat{Y}\})^2\} = (1/N)\sigma_y^2$$

or

$$\sigma_{\hat{y}} = 0.1E\{Y\}$$

$$\sigma_{\hat{y}}^2 = (1/N)\sigma_y^2 = (2/N)\sigma_x^4 = 0.01E\{Y\}^2 = 0.01\sigma_x^4$$

or

$$0.01\sigma_x^4 = (2/N)\sigma_x^4, \qquad \Rightarrow N = 200.$$

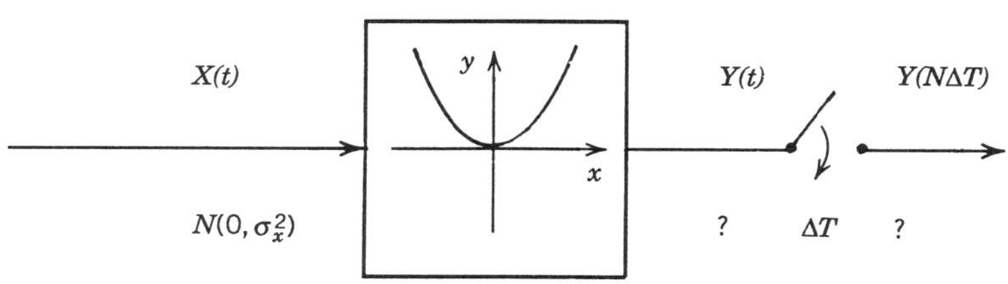

Fig. 2.1 A quadratic impedance

SECOND EXAMPLE Find the running estimates of the output Y and its variance of the previous example. The estimates are calculated as follows to be unbiased:

$$\hat{Y} = (1/N) \sum_{i=1} Y_i,$$

$$\hat{\sigma}_Y^2 = (2/3(N-1)) \sum_{i=1}^{N} \left(Y_i - \hat{Y}\right)^2 = (2/3(N-1)) \sum_{i=1}^{N} \left(Y_i - (1/N) \sum_{j=1}^{N} Y_j\right)^2.$$

Unbiasedness is proven as follows:

$$\begin{aligned}
E\{\hat{\sigma}_Y^2\} &= (2/3(N-1))E\left\{\sum_{i=1}^{N} \left(Y_i - \hat{Y}\right)^2\right\} \\
&= (2/3(N-1))E\left\{\sum_{i=1}^{N} \left(Y_i - (1/N) \sum_{j=1}^{N} Y_j\right)^2\right\} \\
&= (2/3(N-1)) \sum_{i=1}^{N} E\left\{Y_i^2 - (2/N)Y_i \sum_{j=1}^{N} Y_j + (1/N^2) \sum_{k=1}^{N} \sum_{j=1}^{N} Y_k Y_j\right\} \\
&= (2/3(N-1))\left[\sum_{i=1}^{N} E\{Y_i^2\} - 1/N \sum_{k=1}^{N} \sum_{j=1}^{N} E\{Y_k Y_j\}\right] \\
&= (2/3N)E\{Y^2\} = \sigma_Y^2.
\end{aligned}$$

2.2 OPTIMAL ESTIMATES

Optimal estimates may be defined in various ways, depending on the criterion to be optimized. For instance, here are three types of optimal estimates very common to estimation theory:

1. The *most-probable value* is the value that maximizes the a posteriori probability $p(x/Z^m)$ in a probabilistic environment after m random measurements.
2. The *minimum mean- (least-) square value* is the value that minimizes a mean- (least-) square-error criterion.
3. The *minimax value* is the value at the median of the interval of the domain of values.

The three estimates do not necessarily coincide. They do so only when they are associated with a probability density function symmetric about its mean value (Bayes theorem).

Other optimal estimates may be obtained by maximizing the *likelihood function* $L(Z, x) = \ln[p(Z^m/x)]$, of the m measurements of the parameter x. For equal a priori probabilities this estimate is equivalent to the maximum a posteriori probability estimate, which is the most desirable one.

2.3 ACCEPTABLE ESTIMATES

All estimates, optimal or not, must satisfy the following acceptability conditions:

2.3.1 Unbiased Estimates

If the estimate $\hat{x}$ of a parameter vector x, based on subsequent observations, satisfies the condition $E\{\hat{x}\} = E\{x\} = \bar{x}$, then it is called *unbiased*. Otherwise it may have a positive or negative bias;

$$\eta = E\{\hat{x}\} - x.$$

2.3.2 Consistent Estimates

If the estimate $\hat{x}$ of x converges in probability to x, for example,

$$\lim \text{Prob}\left[\|\hat{x}_n - x\| \geq \epsilon\right] = 0, \qquad \text{for arbitrary small } \epsilon,$$

then the estimate x is called *consistent*. A consistent estimate is asymptotically unbiased.

2.3.3 Efficient Estimates

An *efficient* estimate $\hat{x}$ is the unbiased estimate of x with minimum variance, for example,

$$\sigma_x^2 = E\{\|\hat{x} - x\|^2\} \leq E\{\|\hat{y} - x\|^2\}, \qquad \text{for all other estimates } \hat{y} \text{ of } x.$$

2.3.4 Sufficient Estimates

An estimate $\hat{x}$ is called *sufficient* if it contains all the statistical information in the set of measurements Z^m pertaining to the parameter x to be estimated. Alternatively, if Z^m is a finite sequence of measurements pertaining to the parameter x, with probability density function $p(Z^m, x)$, then $\hat{x}$ is the estimate of x. Any other estimate $\hat{y}$ of x conditioned on $\hat{x}$ does not depend statistically on $\hat{x}$; for example, $p(\hat{y}/\hat{x}) = p(\hat{y})$. Any statistic related to a sufficient estimate is called a *sufficient statistic*.

EXAMPLES OF OPTIMAL ESTIMATES See Figure 2.2:

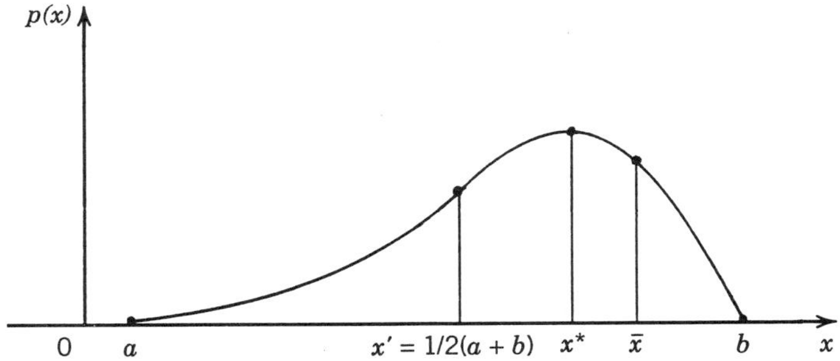

x^*= most probable estimate	$\max_x[p(x)]$
$\bar{x} = E\{x\}$ = minimum risk estimate	$\min_x E\{R(x)\}$
$x' = 1/2(a + b)$ = minimax estimate	median

Fig. 2.2 Optimal estimates

2.4 LEAST-SQUARES ESTIMATION: THE DETERMINISTIC POINT OF VIEW (GAUSS)

PROBLEM Given a set of uneven measurements $Z^m = \{z_1 \cdots z_m\}$ of a quantity x, find the value $\hat{x}$ that makes (LS) minimum, where

$$\mathrm{LS} = \sum_{i=1}^{m} \|\hat{x} - z_i\|^2. \tag{2.4}$$

Assumption:

$$z_i = x + \text{deterministic error}, \qquad i = 1 \cdots m.$$

The answer is obtained under the condition of smoothness of the LS, by differentiation:

$$\partial\,\mathrm{LS}/\partial x = 0 \Rightarrow \sum_{i=1}^{m} (x - z_i) = 0,$$

$$m\hat{x} = \sum_{i=1}^{m} z_i \Rightarrow \hat{x} = (1/m) \sum_{i=1}^{m} z_i.$$

Finally one identifies the unbiased sample mean and variance of the estimate of x, respectively,

$$\hat{x} = (1/m) \sum_{i=1}^{m} z_i = \bar{z}, \tag{2.5}$$

$$\hat{v}_x^2 = (1/(m-1)) \sum_{i=1}^{m} \|z_i - \hat{\bar{z}}\|^2.$$

In the above derivation no assumption was made about the nature of the measurement error. The result yields the best value in the least-square sense. Higher sample moments may be obtained in a similar way:

$$M^n = (1/m) \sum_{i=1}^{m} \|z_i - \bar{z}\|^n. \tag{2.6}$$

EXAMPLE: LEAST-SQUARES PARAMETER ESTIMATION The parameters $q^T = [q_1, q_2, q_3] = [\omega_n^2, 2\zeta\omega_n, g]$ of a second-order system, depicted in Figure 2.3,

$$\begin{bmatrix} dx_1/dt \\ dx_2/dt \end{bmatrix} = \begin{bmatrix} 0 & 1 \\ -\omega_n^2 & -2\zeta\omega_n \end{bmatrix} \begin{bmatrix} x_1 \\ x_2 \end{bmatrix} + \begin{bmatrix} 0 \\ g \end{bmatrix},$$

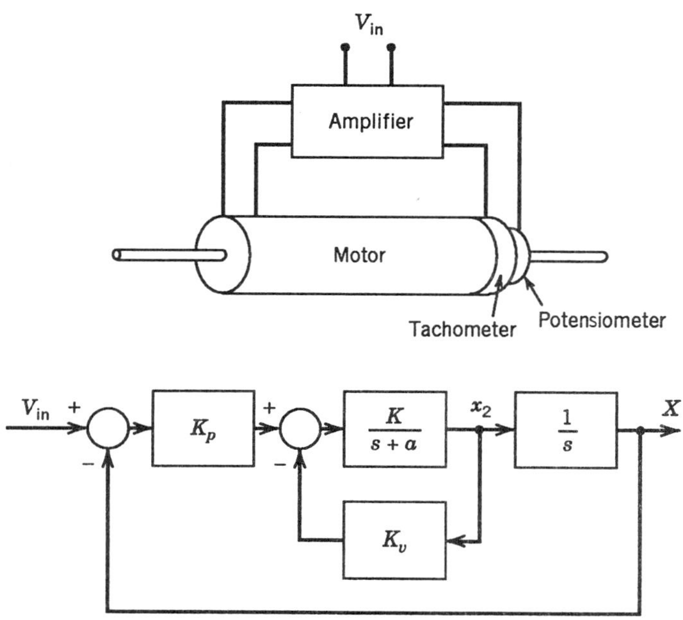

Fig. 2.3 Least-squares example

are to be estimated from time history measurements of displacement x_1, rate x_2, acceleration dx_2/dt, and control u, taken at sampling instants $k = 1 \cdots m$.

Assuming that the measurements of x_1, x_2, and u are precise and those of dx_2/dt are noisy, the measurement equation will read

$$z(k) = dx_2(k)/dt = h(k)^T q + v(k),$$
$$k = 1 \cdots m = \text{number of measurements},$$

where

$$H = \begin{bmatrix} -x_1(1) & -x_2(1) & u(1) \\ -x_1(2) & -x_2(1) & u(2) \\ \cdot & \cdot & \cdot \\ -x_1(m) & -x_2(m) & u(m) \end{bmatrix} = \begin{bmatrix} h(1)^T \\ \cdot \\ \cdot \\ h(m)^T \end{bmatrix}.$$

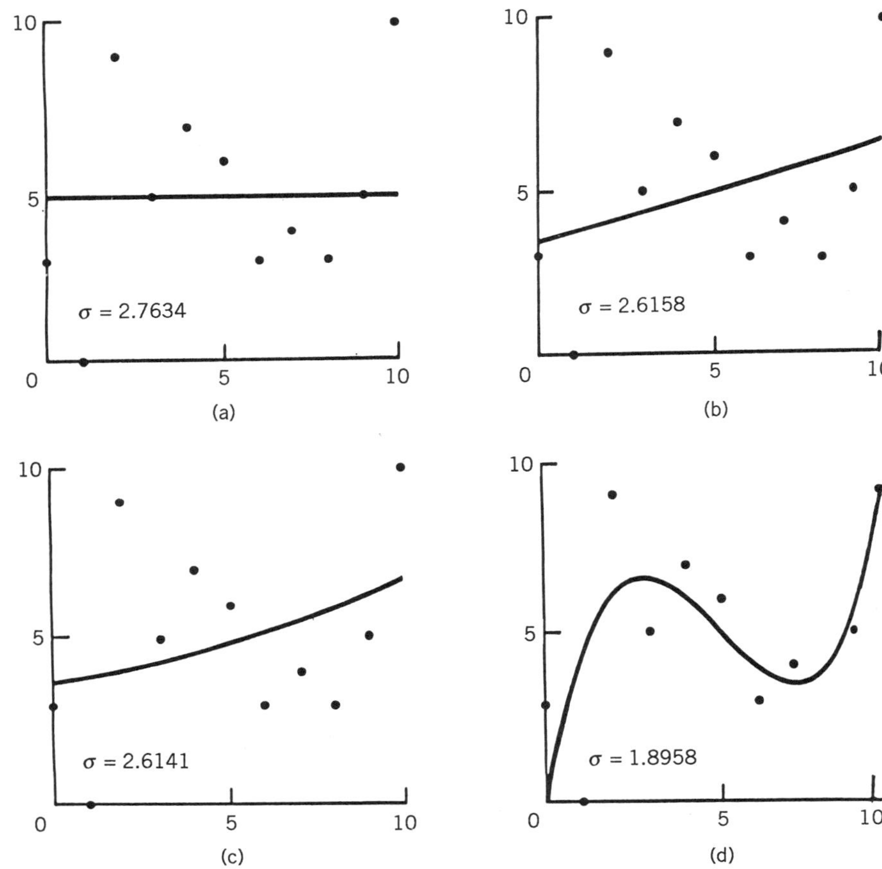

Fig. 2.4 Least-squares curve fitting polynomials : (a) zeroeth degree (b) first degree (c) second degree (d) third degree

The estimates of q are obtained from the estimator [see Eq. (2.30)]:

$$\hat{q} = (H^T H)^{-1} H^T Z^m, \qquad Z^{mT} = [z(1) \cdots z(m)].$$

If a weighting matrix Q is used the above equation is modified to

$$\hat{q} = (H^T Q^{-1} H)^{-1} H^T Q^{-1} Z^m, \qquad Z^{mT} = [z(1) \cdots z(m)].$$

If various-order polynomials were fitted to the measurement data, the plots of Figure 2.4 would result.

2.5 SEQUENTIAL ESTIMATION

A sequential estimator (sample mean) may be obtained by utilizing the previous estimates in conjunction with the last acquired measurement:

$$\hat{x}_m = (1/m) \sum_{i=1}^{m} z_i = (1/m) z_m + (1/m) \sum_{i=1}^{m} z_i = (1/m) z_m$$

$$+ (m-1)/m \left[1/m - 1 \sum_{i=1}^{m-1} z_i \right] = (1/m) z_m + (1 - 1/m) \hat{x}_{m-1}$$

$$= \hat{x}_{m-1} = (1/m)(z_m - \hat{x}_{m-1}). \tag{2.7}$$

This is a form known as stochastic approximation (Robbins and Monroe), which updates every estimate with a weighted error-correction term. Equivalently, the unbiased sample variance is given by

$$\hat{v}_m^2 = 1/(m-1) \sum_{i=1}^{m} (z_i - \hat{x})^2$$

$$= 1/(m-1)(z_m - \hat{x})^2 + 1/(m-1) \sum_{i=1}^{m-1} (z_i - \hat{x})^2$$

$$= 1/(m-1)(z_m - \hat{x})^2 + [1 - 1/(m-1)] \left\{ 1/(m-2) \sum_{i=1}^{m-1} (z_i - \hat{x})^2 \right\}$$

$$= \hat{v}_{m-1}^2 + 1/(m-1) \left[(z_m - \hat{x})^2 - \hat{v}_{m-1}^2 \right].$$

The approach will be demonstrated by the following example: Given a box of

$N = 100$ resistors, of unknown nominal value x, to be found by sequential sample measurements z_i given in Table 2.1,

$$\text{sample mean} = x_{100} = 9.420 + 1/100(10.069 - 9.420) = 9.427,$$

$$\text{variance} = v_{100}^2 = 1.1067 + 1/99(0.642^2 - 1.1067) = 1.1109,$$

where the last measurement was

$$z_{100} = 10.069.$$

See Figure 2.5.

EXAMPLE

TABLE 2.1 Resistance Measurements for Prototype Problem

k	z_k	k	z_k	k	z_k	k	z_k	k	z_k
1	8.835	21	10.924	41	8.046	61	11.203	81	9.875
2	11.896	22	11.506	42	9.082	62	7.815	82	11.123
3	8.283	23	9.486	43	9.978	63	9.559	83	10.723
4	9.986	24	8.754	44	9.237	64	8.859	84	8.013
5	9.011	25	9.843	45	11.120	65	9.189	85	9.138
6	8.148	26	7.647	46	10.318	66	8.348	86	10.311
7	9.115	27	10.819	47	11.462	67	10.416	87	8.400
8	10.123	28	8.707	48	10.256	68	8.703	88	10.761
9	9.531	29	10.134	49	8.822	69	8.620	89	10.081
10	8.940	30	9.365	50	7.945	70	10.622	90	10.635
11	9.623	31	9.915	51	9.115	71	10.925	91	9.043
12	9.044	32	7.484	52	9.589	72	8.381	92	8.077
13	8.347	33	11.598	53	9.757	73	9.174	93	9.924
14	9.169	34	10.503	54	8.279	74	8.583	94	9.172
15	9.938	35	9.020	55	8.368	75	9.574	95	10.788
16	9.027	36	9.589	56	8.720	76	9.758	96	7.838
17	10.824	37	8.908	57	9.776	77	8.746	97	8.596
18	9.551	38	7.472	58	10.049	78	9.973	98	9.545
19	11.012	39	8.293	59	8.702	79	8.741	99	10.413
20	8.282	40	7.283	60	10.421	80	7.911	100	10.069

Sample mean $= \bar{z} = (1/100)\Sigma_{i=1}^{100} z_i = 9.426$.
Sample standard deviation $= s = [(1/100)\Sigma_{i=1}^{100}(z_i - \bar{z})^2]^{1/2} = 1.054$.

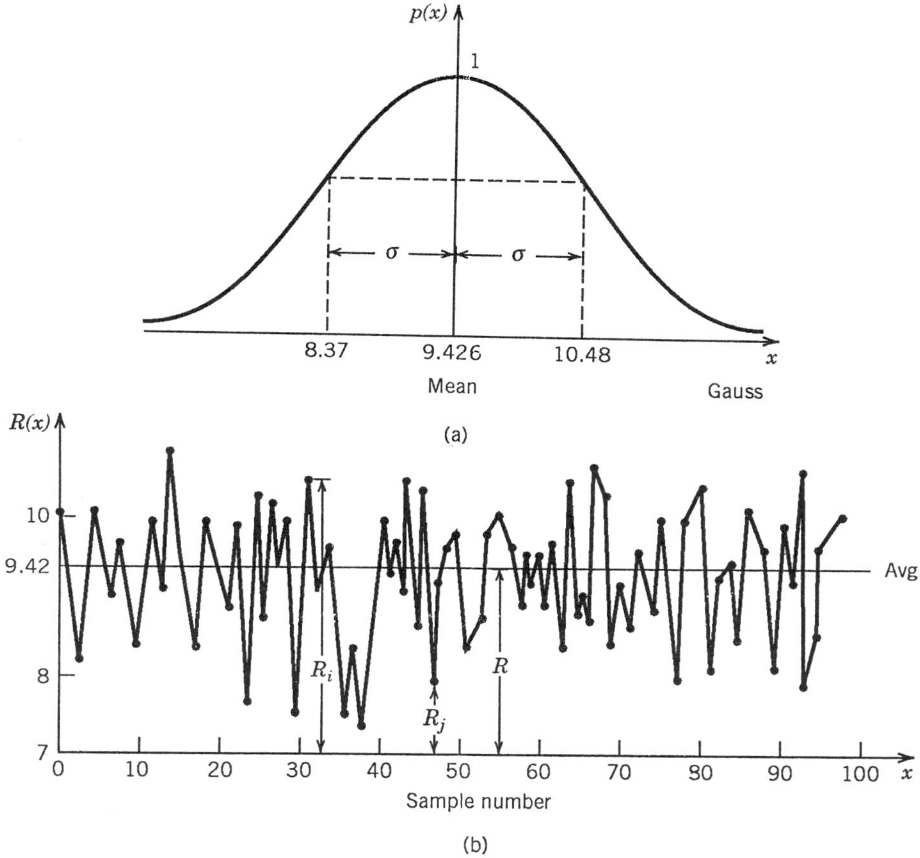

Fig. 2.5 Resistance measurements. (a) Probabilistic (b) Deterministic

2.6 MEAN-SQUARE ESTIMATION: PROBABILISTIC POINT OF VIEW

The probabilistic point of view assumes that the error v_i between measurements and the actual parameter vector x to be estimated is a random variable, which makes the measurements themselves Z^m, as well as the estimates of the parameter vector x, random variables. Define $\tilde{x}$ as the estimation error, with a density function $p(\tilde{x})$, conditional density function $p(\tilde{x}/Z^m)$, its moments $E\{x^n\}$, and *entropy function*

$$H(x) = \int p(x) \ln p(x)\, dx. \tag{2.8}$$

A theorem that interprets the least-squares fit of the estimate $\hat{x}$ to the random measurements in a probabilistic manner is the following.

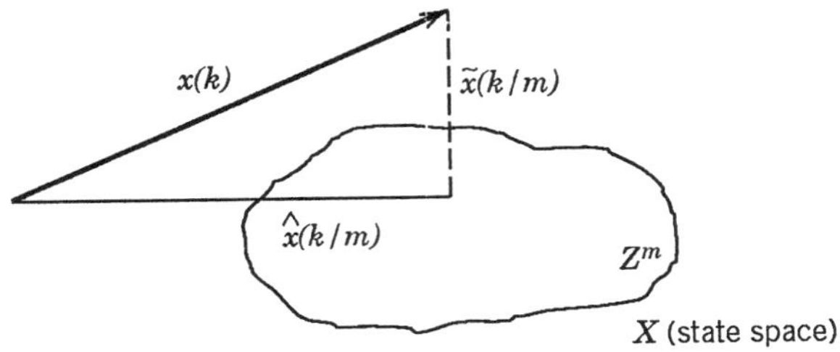

Fig. 2.6 Orthogonal projection

Theorem 2.1 (Orthogonal-Projection Theorem) Let $\hat{x}(k/m) \in Z^m$ be the orthogonal projection of the state $x(k) \in X$, the state space, on the subspace Z^m. Then $\hat{x}(k/m)$ is the best estimate of $x(k)$ on Z^m, and its normal component $\tilde{x}(k/m)$ is orthogonal to Z^m and accounts for the estimation error. See Figure 2.6. The orthogonality condition implies that

1. $\hat{x}(k/m) \in Z^m$.
2. $\tilde{x}(k/m) = \hat{x}(k/m) - x(k)$ is orthogonal to all vectors in Z^m.
3. $\hat{x}(k/m)$ is unique in the Hilbert space X.
4. $\hat{x}(k/m)$ minimizes the mean-square-error criterion $E\{\|\tilde{x}(k/m)\|^2\}$ and the error has minimum covariance, $E\{\tilde{x}(k/m)\tilde{x}^T(k/m)\} = \min$.

Proof

1. The orthogonal projection of $x(k)$ on Z^m is always an element of Z^m.
2. The error $\tilde{x}(k/m)$ is orthogonal on the subspace Z^m; therefore it is orthogonal to any vector $z \in Z^m$; for example, $E\{\langle \tilde{x}(k/m), z \rangle\} = 0$, where $\langle \cdot , \cdot \rangle$ is the inner product symbol.
3. The estimate $\hat{x}(k/m)$ is unique in the Hilbert space X. If not, let vector $b = x - a$ be orthogonal to any $z \in Z^m$, where $\hat{x}(k/m) \neq a \in Z^m$. Then

$$E\{\langle z, b \rangle\} = 0 = E\{\langle z, x \rangle\} - E\{\langle z, a \rangle\} = E\{\langle z, \hat{x} \rangle\} - E\{\langle z, a \rangle\},$$

for all $z \in Z^m$ by Property 2 above. This implies that $\hat{x} \equiv a$. Q.E.D.

4. This property is established by considering an arbitrary z as above:

$$E\{\|x - z\|^2\} = E\{\|x - \hat{x} + \hat{x} - z\|^2\} = E\{\|\hat{x} - x\|^2\} + E\{\|\hat{x} - z\|^2\}$$
$$> E\{\|x - \hat{x}\|^2\} = E\{\|\tilde{x}(k/m)\|^2\} = \min, \qquad \forall\, z \in Z^m.$$

Minimization implies through differentiation of the *mean-square error* that the best estimate $\hat{x}(k/m) = E\{x/Z^m\}$.

Furthermore, $\hat{x}(k/m)$ is the minimum variance estimator by the following argument. For an arbitrary $z \in Z^m$ the error covariance is

$$\begin{aligned}
&E\{(z-x)(z-x)^T\} \\
&\quad = E\{(z-\hat{x}+\hat{x}-x)(z-\hat{x}+\hat{x}-x)^T\} \\
&\quad = E\{(z-\hat{x})(z-\hat{x})^T\} + E\{(x-\hat{x})(z-\hat{x})^T\} \\
&\qquad + E\{(z-\hat{x})(x-\hat{x})^T)\} + E\{(x-\hat{x})(x-\hat{x})^T\} \\
&\quad = E\{(z-\hat{x})(z-\hat{x})^T\} + E\{\tilde{x}\tilde{x}^T\} \geq E\{\tilde{x}\tilde{x}^T\}, \qquad \forall\, z \in Z^m.
\end{aligned}$$

The last two expressions were obtained by invoking Properties 2 and 4 of the theorem, for $\tilde{x} = \hat{x} - x$ and $z - \hat{x}$, respectively. The last relation establishes the minimum variance or *efficient estimator* for the orthogonal-projection theorem.

The orthogonal-projection theorem provides a *minimum mean-square-error* (MMSE) *estimator*, but does not guarantee a most-probable estimator, which may be the most desirable form. This will be established with the Bayes theorem, which will be proven next. It will be shown that the minimum mean-square-error (MMSE) estimator is equivalent to the least-squares estimator (LSE), regardless of the point of view taken in the description of the error.

The value of the MMSE estimator is

$$\hat{x}(k/m) = E\{x(k)/Z^m\}. \tag{2.9}$$

2.7 BAYES THEOREM: MOST PROBABLE VALUE

This theorem yields the same expression as the orthogonal-projection theorem. However, because of its stricter conditions, it results in a much stronger estimator, which in addition to the minimum mean-square estimate yields the most-probable value, based on the measurements

$$Z^m = \{z_1, z_2 \cdots z_m\}. \tag{2.10}$$

The following additional assumptions are made in order to prove the Bayes theorem. Define a loss function $L(\tilde{x})$ that satisfies (see Figure 2.7)

$$\begin{aligned}
&\text{Monotonicity:} \quad L(\tilde{x}_2) > L(\tilde{x}_1), \qquad \text{if } \tilde{x}_2 > \tilde{x}_1, \\
&\text{Symmetry:} \quad L(\tilde{x}) = L(-\tilde{x}).
\end{aligned} \tag{2.11}$$

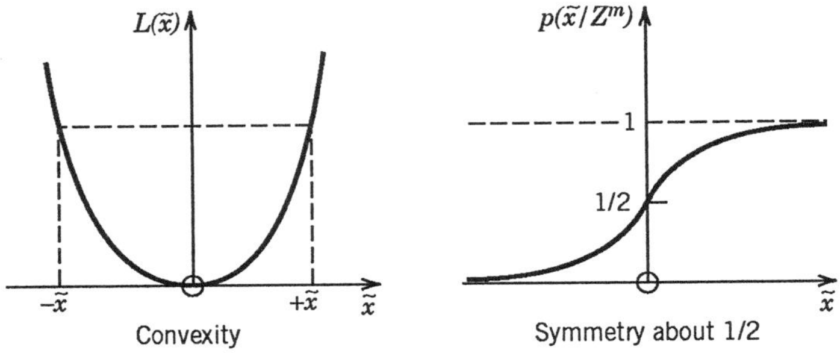

Fig. 2.7 Convexity and symmetry

Define the risk function $R(\tilde{x})$ and the conditional risk function $R(\tilde{x}/Z^m)$, respectively,

$$R(\tilde{x}) = E\{L(\tilde{x})\},$$

$$R(\tilde{x}/Z^m) = E\{L(\tilde{x})/Z^m\}. \tag{2.12}$$

The value of the estimate $\hat{x}$ that minimizes $R(\tilde{x}/Z^m)$ for all $z \in Z^m$ also minimizes $R(\tilde{x})$. Therefore, the conditional risk function will be considered for optimality.

Theorem 2.2 (Bayes Theorem) Assuming that $L(\tilde{x})$ satisfies the monotonicity and symmetry properties mentioned above, and that the conditional distribution function $P(\tilde{x}/Z^m)$ of the random variable $\tilde{x} = \hat{x} - x$ is (a) symmetric about the mean $\bar{x}$, for example

$$P[(\tilde{x} - \bar{x})/Z^m] = 1 - P[(\bar{x} - \tilde{x})/\hat{Z}^m],$$

or (b), convex for $\tilde{x}_1 \leq \tilde{x}_2$, for example,

$$P[\lambda \tilde{x}_1 + (1 - \lambda)\tilde{x}_2/Z^m] \leq \lambda P[\tilde{x}_1/Z^m] + (1 - \lambda)P[\tilde{x}_2/Z^m], \quad \text{for } 0 \leq \lambda \leq 1,$$

the random variable $\hat{x}$ that minimizes the conditional risk $R(\tilde{x}/Z^m)$ of Eq. (2.12) is the *most-probable value* and is given as the conditional expectation of x:

$$\hat{x} = E\{x/Z^m\} = \hat{x}(k/m). \tag{2.13}$$

The same results may be obtained by replacing (b) of the theorem with the convexity of the loss function $L(\tilde{x})$.

In the case of normal distribution functions

$$P(y) = \int_{-\infty}^{y} \frac{1}{\sqrt{2\pi\sigma^2}} \exp\left(-\frac{1}{2}\frac{(x-\bar{x})^2}{\sigma^2}\right) dx,$$

with $\bar{x} = E\{x\}$, $\sigma^2 = E\{(x-\bar{x})^2\}$, the conditions on $L(\tilde{x})$ may be relaxed, and nonsymmetric functions may be considered. The theorem implies that in addition to the most-probable value, the estimator is also MMSE and minimax, as Eq. (2.10) indicates. The conditions on the distribution functions are very difficult to verify in practice, which renders the theorem hard to apply.

Proof: Minimization of $R(\tilde{x}) = E_z\{E_{x/z}[L(\tilde{x})/Z^m\} = E_z\{R(\tilde{x}/Z^m)\}$ is implied by minimization of $R(\tilde{x}/Z^m)$ for all $z \in Z^m$. If conditions (a) and (b) of the theorem are satisfied, and the loss function is monotonic and symmetric, then

$$\min_{\hat{x}} R(\tilde{x}/Z^m) = \min_{\hat{x}} E\{L(\hat{x}-x)/Z^m\} \Leftrightarrow \partial/\partial\hat{x}\, E\{L(\hat{x}-x)/Z^m\}$$

$$= E\{\partial L/\partial\hat{x}\} = 0$$

or

$$\hat{x} = E\{x/Z^m\}. \qquad \text{Q.E.D.}$$

In the special case that $L(\hat{x}-x) = (\hat{x}-x)^2$, the squared-error criterion, minimization of the loss function L yields directly the result of Eq. (2.13). The Bayes theorem establishes the conditions for equivalence of the minimum mean-square- (MMSE) error estimator and the most-probable (maximum a posteriori probability) estimator. In the case that the Bayes theorem conditions are not satisfied, the MMSE estimator is still optimal with respect to the MSE criterion, but is *not* the most-probable value estimator anymore.

2.8 OTHER OPTIMAL ESTIMATORS: MAXIMUM LIKELIHOOD

The maximum-likelihood estimator is based on the Bayes rule; for example,

$$p(x/Z^m) = p(Z^m/x)p(x)/p(Z^m). \tag{2.14}$$

The density $p(Z^m/x)$ is defined as the likelihood density. Its natural logarithm is defined as the likelihood function, and may serve to obtain an efficient estimator for linear systems equivalent to the most-probable estimator:

$$L(Z^m, x) = \ln[p(Z^m/x)]. \tag{2.15}$$

The likelihood function $L(Z^m, x)$ is monotonically increasing with $p(Z^m/x)$. Therefore for uniformly distributed a priori probability densities $p(x)$, minimization of the likelihood function implies minimization of the a posteriori probabilities $p(x/Z^m)$, which yields the most probable estimates.

EXAMPLE

As an example, consider the case of estimating an n-dimensional vector x from an m-dimensional vector of measurements z, related by

$$z = Ax + v,$$

where v is a random m-dimensional noise vector with a Gaussian density function $p(v) \sim N[0, \Lambda]$, independent of x, and A an $m \times n$ constant matrix. The likelihood function now is a function of $\tilde{x}$, namely, $L(\tilde{x})$. The most likely estimate of x is obtained as follows:

$$p(z/x) = p(z, x)/p(x) = p(x, v)/p(x) = p(x)p(v)/p(x) = p(v),$$
$$p(v) = p_v(z - Ax).$$

Because v is Gaussian with zero mean and covariance matrix Λ,

$$p(v) = (2\pi^2|\Lambda|)^{-1/2} \exp\left[-\tfrac{1}{2}v^T\Lambda^{-1}v\right] = C \exp\left\{-\tfrac{1}{2}\|z - Ax\|^2_{\Lambda^{-1}}\right]. \tag{2.16}$$

The *likelihood function* is

$$L(\tilde{x}) = \ln C - \tfrac{1}{2}\|z - A\hat{x}\|^2_{\Lambda^{-1}}.$$

The *maximum* is obtained by

$$\partial L/\partial x = A^T\Lambda^{-1}(z - A\hat{x}) = 0,$$

or, if $m > n$,

$$\hat{x} = (A^T\Lambda^{-1}A)^{-1}A^T\Lambda^{-1}. \tag{2.17}$$

Asymptotically, the density of a maximum-likelihood estimator approaches a normal density, with mean x and variance $1/mI$, where $I = E\{|\partial L(Z^m, \tilde{x})|^2\}$, known as the Rao–Cramér lower bound.

The *maximum-likelihood estimator* is extremely popular for parameter estimation in linear systems and has been investigated by many researchers, such as Sorenson (1980). However, no further discussion will be presented here. The *minimum mean-square-error estimator* will be exclusively considered in this work.

2.9 LINEAR MINIMUM MEAN-SQUARE-ERROR ESTIMATOR: VECTOR CASE

Consider the vector-matrix relation

$$z = Ax + v \tag{2.18}$$

of the m-dimensional vector z of observations, with the n-dimensional vector x of unknown parameters, and an m-dimensional error (noise) random vector v, with zero mean, $E\{v\} = 0$, and finite variance, $\sigma^2 = E\{v^2\} < \infty$; A is an $m \times n$ nonsingular matrix with known elements. Define a *mean-square function* (risk function) to be minimized:

$$R(\tilde{x}) = E\{\|z - A\hat{x}\|^2\} = E_z\{R(\tilde{x}/z)\}. \tag{2.19}$$

If the minimization is over all values of z, as in the case of the Bayes theorem, this is equivalent to minimizing $R(\tilde{x}/z)$:

$$R(\tilde{x}/z) = E\{\|z - A\hat{x}\|^2/z\} = \|z - A\hat{x}\|^2. \tag{2.20}$$

Equation (2.20) establishes the equivalence of the least-squares approach with the minimum mean-square method. From now on the two will be treated indistinguishably.

Case 1 $m \geq n$. The minimization of Eq. (2.20) yields

$$\partial R/\partial x = 0 \Rightarrow A^T(z - A\hat{x}) = 0.$$

Because $m \geq n$, the matrix A^TA is a $n \times n$ nonsingular positive-definite matrix which always has an inverse. Therefore, solving for x,

$$\hat{x} = (A^TA)^{-1}A^Tz, \tag{2.21}$$

which is the minimum mean-square-error (MMSE) estimator of x.

Case 2 $m = n$. For $m = n$,

$$\hat{x} = A^{-1}z. \tag{2.22}$$

Case 3 $m < n$. In this case A^TA is singular and its inverse does not exist. A pseudoinverse $(A^TA)^{\#}$ may be used instead:

$$\hat{x} = (A^TA)^{\#} A^T z. \tag{2.23}$$

One of the most popular pseudoinverses is the Moore–Penrose generalized inverse matrix of rank r, defined to yield the best fit to the data as

$$(A^TA)^{\#} \triangleq C^T[CC^T]^{-1}[B^TB]^{-1}B^T, \tag{2.24}$$

where (A^TA) is decomposed into two maximum rank r matrices B and C,

$$(A^TA)^{[m\times n]} = B^{[m\times r]}C^{[r\times n]}.$$

An obvious criterion to check the dimension of the vector x, if it is unknown, is to test when the matrix (A^TA) stops being singular. The dimension n is then recorded.

If a weighted risk function is used instead of the risk of Eq. (2.19),

$$R(\hat{x}) = E\{(z - A\hat{x})^T Q^{-1}(z - A\hat{x})\}, \tag{2.25}$$

where $Q = Q^T > 0$, the MMSE estimate is given by

$$\hat{x} = (A^TQ^{-1}A)^{-1} A^TQ^{-1}z \tag{2.26}$$

The MMSE estimator is *unbiased*;

$$E\{\hat{x} - x\} = (A^TQ^{-1}A)^{-1}A^TQ^{-1}AE\{\hat{x} - x\} + (A^TQ^{-1}A)^{-1}A^TQ^{-1},$$

$$E\{v\} = 0.$$

The covariance matrix of the MMSE is

$$E\{(\hat{x} - x)(\hat{x} - x)^T\} = \left[(A^TQ^{-1}A)^{-1}A^TQ^{-1}\right]\Lambda\left[(A^TQ^{-1}A)^{-1}A^TQ^{-1}\right]^T,$$

where $\Lambda = E\{vv^T\}$, the noise covariance. In the case that $Q = \Lambda$ is selected, the covariance matrix is reduced to

$$E_\Lambda\{(\hat{x} - x)(\hat{x} - x)^T\} = [A^T\Lambda^{-1}A]^{-1}, \tag{2.27}$$

and the MMSE estimator,

$$\hat{x} = (A^{T\Lambda}\Lambda^{-1}A)^{-1}A^T\Lambda^{-1}z, \tag{2.28}$$

is the *minimum variance estimator*,

$$E_\Lambda\{(\hat{x} - x)(\hat{x} - x)^T\} < E_Q\{(\hat{x} - x)(\hat{x} - x)^T\}, \qquad \forall\, Q \neq \Lambda.$$

An illustrative example demonstrates the method of minimum mean-square-error estimation.

AN ILLUSTRATIVE EXAMPLE A resistive electric circuit with three nodes is fed by three fixed but unknown current sources, represented by a three-dimensional vector x. Four measurements of the voltage at node 1 have been obtained, and they are represented by a four-dimensional vector z. The measurements z are related to the current sources x, through a matrix A of various combinations of resistive values and are corrupted with zero-mean finite variance noise v. Analytically,

$$z = Ax + v,$$

where

$$z = \begin{bmatrix} 10 \\ 23 \\ 31 \\ 12 \end{bmatrix}, \qquad A = \begin{bmatrix} 4 & 2 & 1 \\ 3 & 1 & 0 \\ 0 & 4 & 1 \\ 3 & 4 & 1 \end{bmatrix},$$

$$A^T = \begin{bmatrix} 4 & 3 & 0 & 3 \\ 2 & 1 & 4 & 4 \\ 1 & 0 & 1 & 1 \end{bmatrix},$$

$$(A^TA) = \begin{bmatrix} 34 & 23 & 7 \\ 23 & 37 & 10 \\ 7 & 10 & 3 \end{bmatrix},$$

$$(A^TA)^{-1} = \frac{1}{194}\begin{bmatrix} 11 & 1 & -29 \\ 1 & 53 & -179 \\ -29 & -179 & 729 \end{bmatrix},$$

$$(A^TA)^{-1}A^T = \frac{1}{194}\begin{bmatrix} 17 & 34 & -25 & 8 \\ -69 & 56 & 33 & 36 \\ 255 & -226 & 13 & -74 \end{bmatrix},$$

and the MMSE estimate is

$$\hat{x} = (A^TA)^{-1}A^Tz = \frac{1}{194}\begin{bmatrix} 17 & 34 & -25 & 8 \\ -69 & 56 & 33 & 36 \\ 255 & -266 & 13 & -74 \end{bmatrix}\begin{bmatrix} 10 \\ 23 \\ 31 \\ 12 \end{bmatrix} = \begin{bmatrix} 9.318 \\ 10.589 \\ -20.888 \end{bmatrix}.$$

The result is consistent with the true values assumed.

2.10 SEQUENTIAL MMSE ESTIMATION

Let us consider now the problem of estimating an n-dimensional parameter vector x from a sequence of scalar measurements z_i, corrupted with random noise v_i, where i is the order of appearance in the sequence:

$$z_1 = a_1^T x + v_1$$
$$\cdots$$
$$z_m = a_m^T x + v_m,$$

or in shorthand notation

$$Z^m = A_m x + V^m. \tag{2.29}$$

In the above, $Z^{mT} = \{z_1 \ \cdots \ z_m\}$ is the sequence of m measurements, $A_m^T = [a_1 \ \cdots \ a_m]$ is an $n \times m$ matrix of known coefficients a_i, and V^m is the vector of stacked noise random variables v_i, with zero mean and finite variance:

$$E\{v_i\} = 0, \qquad E\{v_i^2\} < \infty, \qquad \forall\, i.$$

According to the derivation in the previous section, the MMSE estimator after m measurements is given:

$$\hat{x}_m = \left(A_m^T A_m\right)^{-1} A_m^T Z^m, \qquad \text{for } m \geq n. \tag{2.30}$$

When a new measurement z_{m+1} is made,

$$z_{m+1} = a_{m+1}^T x + v_{m+1}, \tag{2.31}$$

one would like to obtain an updated estimate $\hat{x}_{m+1}$ of x without repeating the whole process, using the previous estimate $\hat{x}_m$. This may be obtained by the following procedure. Partition the new matrix-vector measurement equation as follows:

$$Z^{m+1} = \begin{bmatrix} Z^m \\ \hline z_{m+1} \end{bmatrix} = \begin{bmatrix} A_m \\ \hline a_{m+1}^T \end{bmatrix} x + \begin{bmatrix} V^m \\ \hline v_{m+1} \end{bmatrix} = A_{m+1} x + V^{m+1}.$$

The MMSE estimator after $m + 1$ measurements is

$$\hat{x}_{m+1} = \left(A_{m+1}^T A_{m+1}\right)^{-1} A_{m+1}^T Z^{m+1}. \tag{2.32}$$

Now define the matrix P_m such that

$$P_m^{-1} \triangleq A_m^T A_m;$$

then

$$P_{m+1}^{-1} = A_{m+1}^T A_{m+1} = \left[A_m^T | a_{m+1} \right] \left[A_m^T | a_{m+1} \right]^T$$
$$= A_m^T A_m + a_{m+1} a_{m+1}^T = P_m^{-1} + a_{m+1} a_{m+1}^T.$$

Using the *matrix inversion lemma* to invert the recursive relation,

$$P_{m+1} = P_m - P_m a_{m+1} \left(a_{m+1}^T P_m a_{m+1} + 1 \right)^{-1} a_{m+1}^T P_m. \tag{2.33}$$

Substituting this result back to Eq. (2.32), and identifying that

$$A_{m+1}^T Z^{m+1} = A_m^T Z^m + a_{m+1} z_{m+1},$$

one obtains, in conjunction with Eq. (2.33), the *recursive least-squares estimator*.

$$\hat{x}_{m+1} = \hat{x}_m + P_m a_{m+1} \left(a_{m+1}^T P_m a_{m+1} + 1 \right)^{-1} \left(z_{m+1} - a_{m+1}^T \hat{x}_m \right), \tag{2.34}$$
$$P_{m+1} = P_m - P_m a_{m+1} \left(a_{m+1}^T P_m a_{m+1} + 1 \right)^{-1} a_{m+1}^T P_m.$$

The estimation error is given by

$$\tilde{x}_m = \hat{x}_m - x = \left(A_m^T A_m \right)^{-1} A_m^T Z^m - \left(A_m^T A_m \right)^{-1} A_m^T (Z^m - V^m)$$
$$= \left(A_m^T A_m \right)^{-1} A_m^T V^m.$$

It is obvious that the estimator is *unbiased*, because

$$E\{ \tilde{x}_m \} = E\left\{ \left(A_m^T A_m \right)^{-1} A_m^T V^m \right\} = \left(A_m^T A_m \right)^{-1} A_m^T E\{ V^m \} = 0.$$

The covariance matrix of x_m is given by

$$\tilde{P}_m = E\{ \tilde{x}_m \tilde{x}_m^T \} = E\left\{ \left(A_m^T A_m \right)^{-1} A_m^T V^m V^{mT} A_m \left(A_m^T A_m \right)^{-1} \right\}$$
$$= \left(A_m^T A_m \right)^{-1} A_m^T E\{ V^m V^{mT} \} A_m \left(A_m^T A_m \right)^{-1} = \sigma^2 P_m. \tag{2.35}$$

It was assumed that the noise variables are *independent and identically distributed* (i.i.d.), such that

$$E\{ V^m V^{mT} \} = \Lambda = \sigma^2 I.$$

If the risk function is weighted by a diagonal matrix $Q_m = \text{diag}|q_i|$, as in

Eq. (2.25), the estimator x_{m+1} and the matrix P_{m+1} become

$$\hat{x}_{m+1} = \hat{x}_m + P_m a_{m+1}\left(a_{m+1}^T P_m a_{m+1} + q_{m+1}\right)^{-1}\left(z_{m+1} - a_{m+1}^T \hat{x}_m\right), \quad (2.36)$$

$$\tilde{P}_{m+1} = \tilde{P}_m - \tilde{P}_m a_{m+1}\left(a_{m+1}^T \tilde{P}_m a_{m+1} + q_{m+1}\right)^{-1} a_{m+1}^T \tilde{P}_m.$$

In the case that $Q = \Lambda$, the noise covariance matrix, the estimator becomes *minimum variance*, or the *efficient MMSE estimator*, for example,

$$\hat{x}_{m+1} = \hat{x}_m + \tilde{P}_m a_{m+1}\left(a_{m+1}^T \tilde{P}_m a_{m+1} + \sigma^2\right)^{-1}\left(z_{m+1} - a_{m+1}^T \hat{x}_m\right), \quad (2.37)$$

$$\tilde{P}_{m+1} = \tilde{P}_m - \tilde{P}_m a_{m+1}\left(a_{m+1}^T \tilde{P}_m a_{m+1} + \sigma^2\right)^{-1} a_{m+1}^T \tilde{P}_m.$$

The MMSE estimator, Eq. (2.37), is also *consistent*. To prove this let

$$\tilde{P}_m^{-1} = \tilde{P}_{m-1}^{-1} + a_m \Lambda^{-1} a_m^T = \tilde{P}_0^{-1} + \sum_{i=1}^{m} a_i \Lambda^{-1} a_i^T = \tilde{P}_0^{-1} + \sigma^{-2} \sum_{i=1}^{m} a_i a_i^T.$$

For simplicity, let us assume that $a_i = a =$ constant, and

$$\tilde{P}_m^{-1} = \tilde{P}_0^{-1} + \sigma^{-2} m a a^T.$$

Using the matrix inversion lemma,

$$\tilde{P}_m = 1/m\left[\tilde{P}_0^{-1}/m + \sigma^{-2} a a^T\right]^{-1} \approx (1/m)\sigma^{-2}\left[a a^T\right]^{-1} = (1/m)H, \qquad \text{for } m \gg 0,$$

and

$$\lim_{m \to \infty} \tilde{P}_m = \lim_{m \to \infty} \{(1/m)H\} = 0. \quad (2.38)$$

Equation (2.38) suggests that the MMSE estimator has the form of a second-order stochastic approximation.

The MMSE estimator is also *sufficient*, because all the statistical information generated by the measurement sequence Z^m, and the initial guess of $\hat{x}_0$ and $\tilde{P}_0$, is contained in $\hat{x}_{m+1}$ and $\tilde{P}_{m+1}$.

The MMSE estimator is applicable to the state $x(k) \in X$, of the state space of a dynamic system at time k. The evolution of the MMSE estimator for the case of k changing with time is given in the following.

AN ILLUSTRATIVE EXAMPLE For the MMSE sequential estimator the following example illustrates the method.

In the previous example of estimating the current sources of a resistive electric circuit, their best estimate was obtained after four measurements of the voltage at node 1, with different combinations of the circuit's resistive values. When a new measurement of the voltage arrives, under a new circuit configuration, the sequential MMSE estimator is applied to update the values of the current sources previously obtained. One identifies the values of x_4 and P_4, respectively,

$$P_4 = (A^TA)^{-1} = \begin{bmatrix} 0.057 & 0.050 & -0.150 \\ 0.050 & 0.273 & -0.923 \\ -0.150 & -0.923 & 3.760 \end{bmatrix}, \qquad \hat{x}_4 = \begin{bmatrix} 9.318 \\ 10.589 \\ -20.888 \end{bmatrix}.$$

Considering the new measurement z_5 and vector a_5^T, and assuming that σ_v^2, the noise variance,

$$z_5 = 15, \qquad a_5^T = [2\,2\,1], \quad \text{and} \quad \sigma_v^2 = 1,$$

one may calculate the new MMSE estimate x_5 from the recursive expression of Eq. (2.37) for $m = 4$:

$$\hat{x}_{m+1} = \hat{x}_m + \tilde{P}_m a_{m+1}\left(a_{m+1}^T \tilde{P}_m a_{m+1} + \sigma^2\right)^{-1}\left(z_{m+1} - a_{m+1}^T \hat{x}_m\right),$$
$$\tilde{P}_{m+1} = \tilde{P}_m - \tilde{P}_m a_{m+1}\left(a_{m+1}^T \tilde{P}_m a_{m+1} + \sigma^2\right)^{-1} a_{m+1}^T \tilde{P}_m,$$

where

$$a_5^T P_4 = \begin{bmatrix} 2 & 2 & 1 \end{bmatrix} \begin{bmatrix} 0.057 & 0.050 & -0.150 \\ 0.050 & 0.273 & -0.923 \\ -0.150 & -0.923 & 3.760 \end{bmatrix}$$
$$= \begin{bmatrix} 0.064 & -0.277 & 1.614 \end{bmatrix},$$

$$a_5^T P_4 a_5 + \sigma^2 = \begin{bmatrix} 0.064 & -0.277 & 1.614 \end{bmatrix} \begin{bmatrix} 2 \\ 2 \\ 1 \end{bmatrix} + 1 = 2.188,$$

$$a_5^T \hat{x}_4 = \begin{bmatrix} 2 & 2 & 1 \end{bmatrix} \begin{bmatrix} 9.318 \\ 10.589 \\ -20.888 \end{bmatrix} = 18.926.$$

Then the covariance matrix is

$$P_5 = \begin{bmatrix} 0.057 & 0.050 & -0.150 \\ 0.050 & 0.273 & -0.923 \\ 0.150 & -0.923 & 3.760 \end{bmatrix}$$
$$- \frac{1}{2.188} \begin{bmatrix} 0.064 \\ -0.277 \\ 1.614 \end{bmatrix} \begin{bmatrix} 0.064 & -0.277 & 1.614 \end{bmatrix}$$
$$= \begin{bmatrix} 0.055 & 0.058 & -0.103 \\ 0.058 & 0.238 & -0.719 \\ -0.103 & -0.719 & 2.569 \end{bmatrix},$$

and the estimator is

$$\hat{x}_5 = \begin{bmatrix} 9.318 \\ 10.589 \\ -20.888 \end{bmatrix} + 1/2.188 \begin{bmatrix} 0.064 \\ -0.277 \\ 1.614 \end{bmatrix} [\,15.000 \quad -18.926\,] = \begin{bmatrix} 9.203 \\ 11.086 \\ -23.784 \end{bmatrix}.$$

2.11 NONLINEAR PARAMETER ESTIMATION

In most systems, parameters enter the measurement equation linearly. However, in many cases, a parameter vector x may be related to the measurements in a nonlinear manner:

$$z = f(x) + v, \tag{2.39}$$

where z is an m-dimensional measurement vector, x is the n-dimensional parameter vector, v is a zero-mean finite covariance, $Q = Q^T > 0$ is a random vector, and $f(\cdot)$ is a nonlinear function, twice differentiable with respect to its argument. There exists no exact method to estimate the parameter vector x. The best way to obtain results, in a case like this, is by linearization of the nonlinear function.

As in the case of the Bayes theorem, Eqs. (2.19) and (2.20), the risk function is

$$R(\tilde{x}) = E\{R(\tilde{x}/z)\} = E\{[z - f(x)]^T Q[z - f(x)]\}, \tag{2.40}$$

$$R(\tilde{x}/z) = E\{[z - f(x)]^T Q[z - f(x)]/z\} = [z - f(x)]^T Q^{-1}[z - f(x)].$$

Assuming that there exists an initial estimate $x(0) = x_0$, close to x, and that $f(x)$ is Taylor-series expandable, such that it can be approximated by

$$f(x) \approx f(\hat{x}(k)) + \nabla_x f(\hat{x}(k))^T [x - \hat{x}(k)],$$

the following recursive algorithm yields an approximate MMSE estimate:

$$\begin{aligned} \hat{x}(k) = \hat{x}(k-1) &+ \left[\nabla_x f(\hat{x}(k-1)) Q^{-1} \nabla_x f(\hat{x}(k-1))^T\right]^{-1} \\ &\times \nabla_x f(\hat{x}(k-1)) Q^{-1}[z - f(\hat{x}(k-1))]. \end{aligned} \tag{2.41}$$

PROBLEMS

2.1 Two sensors measure the same unknown parameter x:

$$Z_i = x + V_i, \qquad i = 1, 2,$$

where the random vectors V_1 and V_2 have zero means and variances σ_1^2

and σ_2^2, respectively. The proposed estimator of x is

$$\hat{x} = K_1 Z_1 + K_2 Z_2$$

Determine the gains K_1 and K_2 such that the estimator should be unbiased, and minimize the mean-square estimation error = $E\{(\hat{x} - x)^2\}$.

2.2 A sequence of scalar measurements $\{Z_i\}$, $i = 1, 2 \cdots n$, of a constant but unknown quantity x is given by

$$Z_i = x + V_i,$$

where V_i is an independent identically distributed Gaussian noise $V_i \sim N[0 \cdot \sigma^2]$, $i = 1 \cdots n$. Prove that the estimator of x,

$$\hat{x} = (1/n) \sum^{n} Z_i,$$

is (a) unbiased, (b) consistent.
Note: Use the Chebychev inequality; $\text{Prob}[|\hat{x} - x| > \epsilon] \leq \sigma_x^2/\epsilon^2$.

2.3 Show that the maximum-likelihood estimator is equivalent to the most-probable estimator if the a priori probabilities $p(x)$ are uniformly distributed.

2.4 Derive the sequential mean-square estimator and the covariance matrix propagation formula if the mth estimate is given by

$$\hat{X}_m = \left(A_m^T A_m\right)^{-1} A_m^T Z_m,$$

and the new measurement is

$$Z_{m+1} = A_{m+1}^T X + V_{m+1},$$

where V_i, $i = 1 \cdots m + 1$ is zero mean, i.i.d., and $E\{V_i^2\} = \sigma^2 < \infty$; n is the dimension of vectors a_i, $i = 1 \cdots m + 1$, for which independent and known vector coefficients form the matrices $A_m^T = [a_1 \ \cdots \ a_m]$.

REFERENCES

Aoki, M., *Introduction to Stochastic Systems*, Academic Press, New York, 1967.

Deutz, R., *Estimation Theory*, Prentice Hall, Englewood Cliffs, NJ, 1965.

Doob, J. L., *Stochastic Processes*, Wiley, New York, 1953.

Dynkin, E. B., *Markov Processes*, Academic Press, New York, 1953.

Gelb, A., *Applied Optimal Estimation*, MIT Press, Cambridge, MA, 1974.

Kailath, T., and Frost, D., "An Innovations Approach to Least Squares Estimation, Parts I, II, III, IV," *IEEE Trans. Automatic Control*, Vol. AC-13, No. 6, pp. 655–660 (1968).

Lewis, F. L., *Optimal Estimation: Introduction to Stochastic Control*, Wiley-Interscience, New York, 1986.

Sage, A. P., and Melsa, J. L., *System Identification*, Academic Press, New York, 1971.

Saridis, G. N., *Self-Organizing Controls of Stochastic Systems*, Marcel Dekker, New York, 1977.

Skorokhod, A. V., *Studies in the Theory of Random Processes*, Addison-Wesley, Reading, MA, 1965.

Sorenson, H. W., *Parameter Estimation*, Marcel Dekker, New York, 1980.

Wilde, D., *Optimal Seeking Methods*, Prentice-Hall, New York, 1963.

3

OPTIMAL STATE ESTIMATION

3.1 DISCRETE-TIME OPTIMAL STATE ESTIMATION: THE KALMAN FILTER

In the previous section, various methods were proposed to estimate unknown parameters of a system, fixed in time, from a set of measurements corrupted with noise. These methods can be equally applied to a dynamic system, frozen in time k, with the states $x(k)$ as the unknown parameters. This was the idea when the orthogonal-projection theorem was presented and estimates $\hat{x}(k/m)$ were derived. A method of extending the estimation results to time-varying states will be developed, and an *optimal state estimator* will be produced, for the *discrete-time* case first. This work was originated by Kalman and Bucy for the continuous-time systems that will be presented in the following. The approach will be based on the Orthogonal Projection theorem, which yields the minimum mean-square-error (MMSE) estimator. In the case of noise distributions that are symmetric about the mean, like the Gaussian, and that are invariantly propagated through the system, for example, as in a linear system, the estimator is most probable as well, because it satisfies the conditions of the Bayes theorem. If not, the estimator is still optimal in the mean-square sense.

Consider the following n-dimensional discrete-time system, forming a stochastic process:

$$\begin{aligned} x(k+1) &= F(k)x(k) + G(k)w(k), \qquad x(0), \qquad (3.1) \\ z(k) &= H(k)x(k) + v(k), \end{aligned}$$

where a measurement m vector $z(k)$ is replacing the scalar case previously considered; the coefficient n vector a_m is replaced by the $m \times n$ matrix $H(k)$. The stochastic variables $x(0)$ represent the n-dimensional initial state, $w(k)$ is the r-dimensional process noise, and $v(k)$ is the m-dimensional measurement noise; for any time k they have the following properties:

$$E\{x(0)\} = x_0, \quad E\{(x(0) - x_0)(x(0) - x_0)^T\} = P_0 \tag{3.2}$$

$$E\{w(k)\} = 0, \quad E\{w(k)w(j)^T\} = Q(k)\,\delta_{kj}, \qquad E\{v(k)v(j)^T\} = R(k)\,\delta_{kj},$$

$$E\{v(k)\} = 0, \quad E\{v(k)x(0)^T\} = E\{w(k)x(0)^T\} = E\{w(k)v(j)^T\} = 0.$$

In the above properties, indicate that the stochastic processes are uncorrelated and *independent and identically distributed* (i.i.d), and δ_{kj} is the Kronecker delta, defined $\delta_{kk} = 1$, $\delta_{kj} = 0$ for $k \neq j$, and $\forall\, k, j$. The orthogonal projection (Bayes) theorem, as applied to sequential estimation of the states at time k, and after m measurements, yields

$$\hat{x}(k/m) = E\{x(k)/Z^m\},$$
$$\tilde{x}(k/m) = \hat{x}(k/m) - x(k),$$

and error covariance

$$\tilde{P}(k/m) = E\{\tilde{x}(k/m)\tilde{x}(k/m)^T\}.$$

The equations of propagation of the sequential estimator and its covariance matrix, defined by Eq. (2.37) of the previous section, are

$$\hat{x}_m = \hat{x}_{m-1} + \tilde{P}_{m-1}a_m\left(a_m^T\tilde{P}_{m-1}a_m + \sigma^2\right)^{-1}\left(z_m - a_m^T\hat{x}_{m-1}\right), \qquad x_0, \tag{3.3}$$

$$\tilde{P}_m = \tilde{P}_{m-1} - \tilde{P}_{m-1}a_m\left(a_m^T\tilde{P}_{m-1}a_m + \sigma^2\right)^{-1}a_m\tilde{P}_{m-1}, \qquad P_0.$$

In order to estimate the present state $x(k)$, after k measurements Z^k, replace in Eq. (3.3), $m = k$, $\hat{x}_{m-1} = \hat{x}(k/k-1)$, $\hat{x}_m = \hat{x}(k/k)$, $\tilde{P}_{m-1} = \tilde{P}(k/k-1)$, $\tilde{P}_m = \tilde{P}(k/k)$, and in order to be consistent with the dimensions of the system of Eq. (3.1), a_m^T with the matrix $H(k)$. The MMSE sequential estimator of the state $x(k)$ and its covariance then read

$$\begin{aligned}
\hat{x}(k/k) &= \hat{x}(k/k-1) + \tilde{P}(k/k-1)H(k)^T \\
&\qquad \times\left[H(k)\tilde{P}(k/k-1)H(k)^T + R(k)\right]^{-1}\Big] \\
&\qquad \cdot\left[z(k) - H(k)\hat{x}(k/k-1)\right], \\
\tilde{P}(k/k) &= \tilde{P}(k/k-1) - \tilde{P}(k/k-1)H(k)^T \\
&\qquad \times\left[H(k)\tilde{P}(k/k-1)H(k)^T + R(k)\right]^{-1}H(k)\tilde{P}(k/k-1).
\end{aligned} \tag{3.4}$$

Up to now the state $x(k)$ of the dynamic system was estimated from the measurement equation of Eq. (3.1), as if it were a parameter of the system. In order to consider the dynamic evolution of the state $x(k)$, the estimate of the state $x(k+1)$ must be obtained, using the orthogonal projection (Bayes) theorem, after k measurements, for example, Eq. (2.3):

$$\hat{x}(k+1/k) = E\{F(k)x(k) + G(k)w(k)/Z^k\} = F(k)\hat{x}(k/k). \quad (3.5)$$

Substituting Eq. (3.4) into Eq. (3.5), one obtains the one-step-ahead prediction estimator of the dynamic system of Eq. (3.1):

$$\begin{aligned}\hat{x}(k+1/k) &= F(k)\hat{x}(k/k-1) + F(k)\tilde{P}(k/k-1)H(k)^T \\ &\quad \times\left[H(k)\tilde{P}(k/k-1)H(k)^T + R(k)\right]^{-1} \\ &\quad \cdot[z(k) - H(k)\hat{x}(k/k-1)]. \end{aligned} \quad (3.6)$$

The propagation of the error covariance matrix is obtained after defining the propagation of the error $\tilde{x}(k+1/k) \perp Z^k$:

$$\begin{aligned}\tilde{x}(k+1/k) &= \hat{x}(k+1/k) - x(k+1) \\ &= F(k)\tilde{x}(k/k-1) + F(k)\tilde{P}(k/k-1)H(k)^T \\ &\quad \times\left[H(k)\tilde{P}(k/k-1)H(k)^T + R(k)\right]^{-1} \\ &\quad \times[z(k) - H(k)\hat{x}(k/k-1)] - G(k)w(k). \end{aligned} \quad (3.7)$$

The covariance matrix is

$$\begin{aligned}\tilde{P}(k+1/k) &= E\{\tilde{x}(k+1/k)\tilde{x}(k+1/k)^T\} = F(k)E\{(\hat{x}(k/k) - x(k)) \\ &\quad \cdot(\hat{x}(k/k) - x(k)^T)\}F(k)^T - F(k)E\{(\hat{x}(k/k) - x(k))w(k)^T\}G(k)^T \\ &\quad - G(k)E\{w(k)(\hat{x}(k/k) - x(k))^T\}F(k)^T \\ &\quad + G(k)E\{w(k)w(k)^T\}G(k)^T\} \\ &= F(k)\tilde{P}(k/k)F(k)^T + G(k)Q(k)G(k)^T. \end{aligned} \quad (3.8)$$

Equations (3.7) and (3.8) describe completely the propagation of the MMSE estimate and its covariance, and are useful when digital computers are used for state estimation, when one-step-ahead prediction is needed when computation time is required. The estimator, usually called the Kalman filter (see

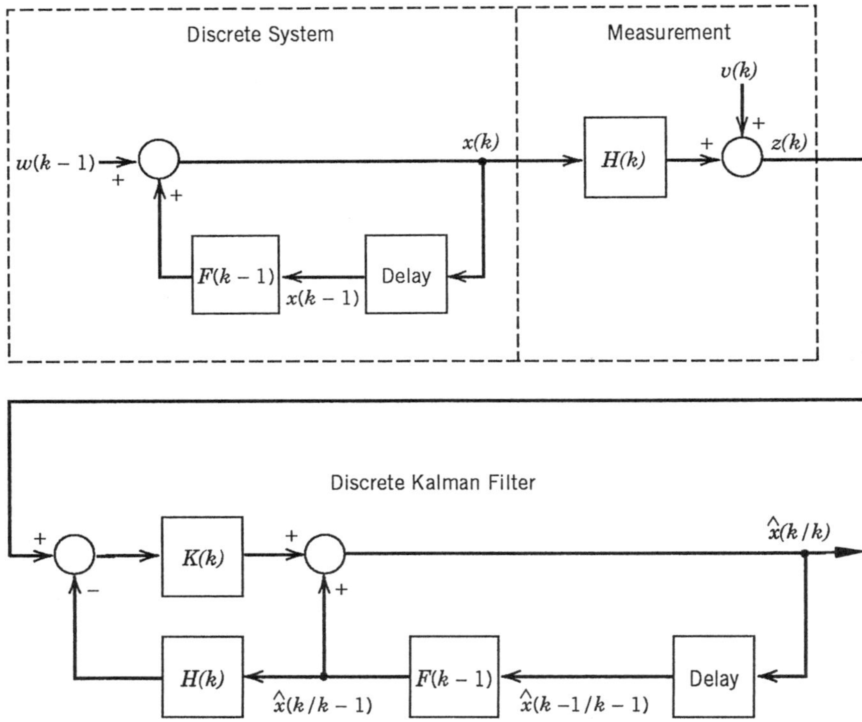

Fig. 3.1 System model and discrete Kalman filter

Figure 3.1), after its originator, is given by

$$\begin{aligned}
\hat{x}(k/k) &= F(k-1)\hat{x}(k-1/k-1) + \tilde{P}(k/k)H(k)^T R(k)^{-1} \\
&\quad \times [z(k) - H(k)F(k-1)\hat{x}(k-1/k-1)], \\
\tilde{P}(k/k) &= \tilde{P}(k/k-1) - \tilde{P}(k/k-1)H(k)^T \\
&\quad \times \left[H(k)\tilde{P}(k/k-1)H(k)^T + R(k)\right]^{-1} H(k)\tilde{P}(k/k-1), \\
\tilde{P}(k/k-1) &= F(k-1)\tilde{P}(k-1/k-1)F(k)^T \\
&\quad + G(k-1)Q(k-1)G(k-1)^T, \qquad (3.9) \\
\hat{x}(0/0) &= E\{x(0)\}, \qquad \tilde{P}(0/0) = E\left\{(\hat{x}(0/0) - x_0)(\hat{x}(0/0) - x_0)^T\right\} = P_0.
\end{aligned}$$

TABLE 3.1 Summary of Discrete Kalman Filter Equations

System model	$x(k) = F(k-1)x(k-1) + G(k-1)w(k-1),$	$w \sim N(0, Q(k))$
Measurement model	$z(k) = H(k)x(k) + v(k), \qquad v(k) \sim N(0, Q(k))$	
Initial conditions	$E\{x(0)\} = \hat{x}_0, \qquad E\{[x(0) - \hat{x}_0][x(0) - \hat{x}_0]^T\} = P_0$	
Other assumptions	$E\{w(k)v(j)\} = 0, \qquad \forall\, k, j$	
State estimate	$\hat{x}(k/k-1) = F(k)\hat{x}(k-1/k-1)$	
Error covariance	$\tilde{P}(k/k-1) = F(k)\tilde{P}(k-1/k-1)F(k-1)^T + GQ(k-1)G^T$	
Estimate update	$\hat{x}(k/k) = \hat{x}(k/k-1) + K(k)[z(k) - H(k)\hat{x}(k/k-1)]$	
Covariance update	$\tilde{P}(k/k) = [I - K(k)H(k)]\tilde{P}(k/k-1)$	
KF gain matrix	$K(k) = \tilde{P}(k/k-1)H(k)^T[H(k)\tilde{P}(k/k-1)H(k)^T + R(k)]^{-1}$	

3.2 PROPERTIES OF THE KALMAN FILTER (KF)

3.2.1 The Kalman Filter is Unbiased

Proof: This is proven by induction:

$$\hat{x}(1/1) = F(0)\hat{x}(0/0) + K(1)[z(1) - H(1)F(0)\hat{x}(0/0)],$$
$$E\{\hat{x}(1/1) - x(1)\} = F(0)E\{\hat{x}(0/0) - x(0)\}$$
$$+K(1)[H(1)F(0)[E\{\hat{x}(0/0) - x(0)\}] = 0,$$

because $E\{\hat{x}(0/0)\} = E\{x(0)\}$. By induction, if $E\{\hat{x}(k-1/k-1)\} = E\{x(k-1)\}$,

$$\begin{aligned} &E\{\hat{x}(k/k) - x(k)\} \\ &\quad = E\{F(k-1)[\hat{x}(k-1/k-1) - x(k)]\} \\ &\qquad + K(k)E\{z(k) - H(k)F(k-1)\hat{x}(k-1/k-1)\} \\ &\quad = F(k-1)E\{\hat{x}(k-1/k-1) - x(k-1)\} \\ &\qquad + H(k)F(k-1)E\{\hat{x}(k-1/k-1) - x(k)\} = 0. \quad \text{Q.E.D.} \end{aligned}$$

In the above derivations it was assumed that the Kalman gain is

$$K(k) = \tilde{P}(k/k-1)H(k)^T\left[H(k)\tilde{P}(k/k-1)H(k)^T + R(k)\right]^{-1} \tag{3.10}$$

3.2.2 The Kalman Filter is Consistent

Proof: Because the asymptotic properties of the filter are investigated, assume that the system coefficients are constant, especially $H(k) = H$. Assume also, without loss of generality, that $R(k) = \sigma^2 I$, where I is the

m-dimensional identity matrix. Then it is obvious that

$$\begin{aligned}\tilde{P}(k/k)^{-1} &= \tilde{P}(k/k-1)^{-1} + H^T R H = \tilde{P}(k/k-1)^{-1} + \sigma^2 H^T H \\ &= \tilde{P}(0/-1)^{-1} + \sigma^2 \sum_{i=1}^{k} H^T H = \tilde{P}(0/-1)^{-1} + k\sigma^2 H^T H.\end{aligned}$$

Inverting the above,

$$\tilde{P}(k/k) = (1/k)\left[\tilde{P}(0/-1)^{-1}/k + \sigma^2 H^T H\right]^{-1} \underset{k\to\infty}{\to} 0. \quad \text{Q.E.D.}$$

3.2.3 The Kalman Filter is Efficient

The MMSE estimator, used to obtain the KF, was shown to be minimum variance. Therefore, the Kalman–Bucy filter is minimum variance, and is consequently efficient.

3.2.4 The Kalman Filter is Sufficient

Because the MMSE estimator was using sufficient statistics, the KF derived from it is also sufficient. Finally, if one needs to calculate a predictor of the state at time q, given k measurements Z^k, it is generated by

$$\hat{x}(q/k) = \prod_{i=1}^{q-1} F(i)\hat{x}(k/k). \tag{3.11}$$

AN ILLUSTRATIVE EXAMPLE OF THE DISCRETE-TIME KALMAN FILTER Consider the first-order discrete-time system corrupted with measurement noise only:

$$\begin{aligned}x(k+1) &= x(k), \qquad x(0), \\ z(k) &= x(k) + v(k),\end{aligned}$$

where $x(0)$ is a random variable with x_0, p_0, $v(k) \sim N(0, r_0)$, $f(k) = h(k) = 1$, $g(k) = q(k) = 0$; the above yield an error variance propagation equation:

$$\begin{aligned}\tilde{P}(k+1/k) &= \tilde{P}(k/k), \qquad \tilde{P}(0/0) = p_0, \\ \tilde{P}(k+1/k+1) &= \tilde{P}(k+1/k) \\ &\quad - \tilde{P}(k+1/k)\left[\tilde{P}(k+1/k) + r_0\right]^{-1}\tilde{P}(k+1/k) \\ &= \tilde{P}(k+1/k)/\left[1 + \tilde{P}(k+1/k)/r_0\right] \\ &= \tilde{P}(k/k)/\left[1 + \tilde{P}(k/k)r_0\right].\end{aligned}$$

Iteratively,

$$\tilde{P}(1/1) = p_0/[1 + p_0/r_0] \cdots \tilde{P}(k/k) = p_0/[1 + kp_0/r_0],$$

and the Kalman estimator

$$\hat{x}(k+1/k+1) = \hat{x}(k/k) + \left[\tilde{P}(k/k)/r_0\right]\left[z(k) - \hat{x}(k/k)\right], \qquad \hat{x}(0/0) = x_0.$$

To show unbiasedness inductively, considering that $E\{\tilde{x}(0/0)\} = 0$,

$$\begin{aligned} &E\{\hat{x}(1/1) - x(1)\} \\ &\quad = E\{\tilde{x}(0/0) + [p_0/[p_0 + r_0]][v(1) - \tilde{x}(0/0)]\} = 0, \\ &E\{\hat{x}(k+1/k+1) - x(k+1)\} \\ &\quad = E\{\tilde{x}(k/k) + [p_0/[r_0 + p_0 k]][v(k) - \tilde{x}(k/k)]\} = 0. \end{aligned}$$

To show consistency, consider that error variance propagation equation

$$\lim_{k \to \infty} \tilde{P}(k/k) = \lim_{k \to \infty} p_0/[1 + kp_0/r_0] = 0.$$

3.3 STOCHASTIC CONTROLLABILITY, OBSERVABILITY, AND STABILITY

The asymptotic behavior of an estimator defines its performance as $k \to \infty$, and requires stability of operation. This makes sense if it is assumed that the coefficients, $F(k) = F$, $G(k) = G$, $H(k) = H$, $Q(k) = Q$, $R(k) = R$, are constant. In order to study the stability of the estimator the following properties must be defined.

3.3.1 Stochastic Controllability

Equation (3.8) was defined as the covariance propagation equation:

$$\tilde{P}(k+1/k) = F(k)\tilde{P}(k/k)F(k)^T + G(k)Q(k)G(k)^T$$

and represents the matrix Riccati equation in the absence of measurements. The equation is linear and for constant coefficients is

$$\tilde{P}(k+1/k) = F\tilde{P}(k/k) + \tilde{P}(k/k)F^T + GQG^T.$$

The system (3.1) is said to be *uniformly completely controllable* if

$$\beta_1 I \le \sum_{i=k-N}^{k-1} \Phi(k, i+1) Q \Phi(k, i+1)^T \le \beta_2 I, \qquad \Phi(k, j) = F^{k-j},$$

for some $N > 0$, and $\beta_1, \beta_2 > 0$.

3.3.2 Stochastic Observability

In the absence of process noise and constant coefficients, system (3.1) is called *uniformly completely observable* if

$$\gamma_1 I \le \sum_{i=k-N}^{k-1} \Phi(i,k)^T H^T R^{-1} H \Phi(i,k) \le \gamma_2 I, \qquad \Phi(k,j) = F^{k-j},$$

for some $N > 0$, and $\gamma_1, \gamma_2 > 0$.

3.3.3 Asymptotic Stability

The estimator of Eq. (3.9), may be written as a forced discrete-time system

$$\hat{x}(k/k) = \left[I - \tilde{P}(k/k)H^T R^{-1} H\right] F\hat{x}(k-1/k-1) + \tilde{P}(k/k)H^T R^{-1} z(k).$$

It is desirable that the unforced estimator be asymptotically stable, for example,

$$\hat{x}(k/k) \to 0, \text{ as } t \to \infty, \text{ for } z(k) = 0, \text{ and any initial condition.}$$

Optimality of the Kalman filter does not guarantee stability. However, the unforced system is asymptotically stable, and the covariance $\tilde{P}(k/k)$ is unique for large k, independent of the value $\tilde{P}(0/0)$, if *stochastic uniform controllability* and *uniform complete observability* are guaranteed, for constant system coefficients. This property is a direct result of Kalman (1964).

3.4 CONTINUOUS-TIME OPTIMAL STATE ESTIMATION: THE KALMAN–BUCY FILTER

The continuous-time MMSE state estimator was derived independently and then combined by Kalman and Bucy, in 1960. The combined results involved the Wiener filter, which was derived earlier, and required a variational approach for its derivation. A simpler approach, which treats the continuous-time case at the limit of the discrete-time case, as the discrete-time interval $\Delta t \to 0$, follows. Assuming the existence of impulse or Dirac delta functions $\delta(t-\tau)$, this approach reduces all the mathematical operations to simple algebraic calculations. The MMSE (Bayes) state estimator is identical to the Kalman–Bucy results as they apply to a linear stochastic system. See Table 3.2.

Consider the following continuous-time system, forming an n-dimensional linear stochastic process:

$$\begin{aligned} dx/dt &= F(t)x(t) + G(t)w(t), \qquad x(0), \\ z(t) &= H(t)x(t) + v(t), \end{aligned} \tag{3.12}$$

TABLE 3.2 Summary of Continuous Kalman–Bucy Filter Equations

System model	$dx/dt = F(t)x(t) + G(t)w(t), \quad w(t) \sim N(0, Q(t))$
Measurement model	$z(t) = H(t)x(t) + v(t), \quad v(t) \sim N(0, R(t))$
Initial conditions	$E\{x(0)\} = \hat{x}_0, \quad E\{[x(0) - \hat{x}_0][x(0) - \hat{x}_0]^T\} = P_0$
Other assumptions	$R(t)^{-1} > 0$ exists $E\{w(t)v(\tau)\} = 0$
State estimate	$d\hat{x}/dt = F(t)\hat{x}(t) + K(t)[z(t) - H(t)\hat{x}(t)]$, $\hat{x}(0) = \hat{x}_0$
Error covariance	$d\tilde{P}/dt = F(t)\tilde{P}(t)\tilde{P}(t)F(t)^T + G(t)Q(t)G(t)^T - K(t)R(t)K(t), \quad \tilde{P}(0) = \tilde{P}_0.$
KB gain matrix	$K(t) = \tilde{P}(t)H(t)^T R(t)^{-1}$

where $x(t)$ is the state of the system at time t, $z(t)$ is the current m-dimensional measurement vector, and $F(t), G(t)$ are $n \times n$ and $n \times r$ matrices of known coefficients, respectively. The stochastic variables $x(0)$, n-dimensional initial state, $w(t)$, r-dimensional process noise, $v(k)$, m-dimensional measurement noise, defined for any $t \in [0, \infty]$, have the following properties:

$$\begin{aligned} E\{x(0)\} &= x_0, & E\{(x(0) - x_0)(x(0) - x_0)^T\} &= P_0, \\ E\{w(t)\} &= 0, & E\{w(t)w(\tau)^T\} &= Q(t)\,\delta(t - \tau), \\ E\{v(t)\} &= 0, & E\{v(t)v(\tau)^T\} &= R(t)\,\delta(t - \tau). \end{aligned} \tag{3.13}$$

The above properties indicate that the stochastic processes are i.i.d. and possess finite means and covariances. In addition,

$$E\{x(0)w(t)^T\} = E\{x(0)v(t)^T\} = E\{w(t)v(\tau)^T\} = 0. \tag{3.14}$$

Time discretization of the above system means that the infinitesimal time interval dt becomes the finite discrete interval Δt, and all the related variables take values at the discrete-time intervals $k\,\Delta t \propto k$. The Dirac delta function $\delta(t)$ becomes a pulse of width Δt and height $1/\Delta t$, corresponding to the Kronecker delta $\delta_{k,k+1}$, at times $k\,\Delta t$ and $(k+1)\,\Delta t$, respectively. See Figure 3.2. Obviously,

$$\lim_{\Delta t \to \infty} \delta_{k,k+1} = \delta(t). \tag{3.15}$$

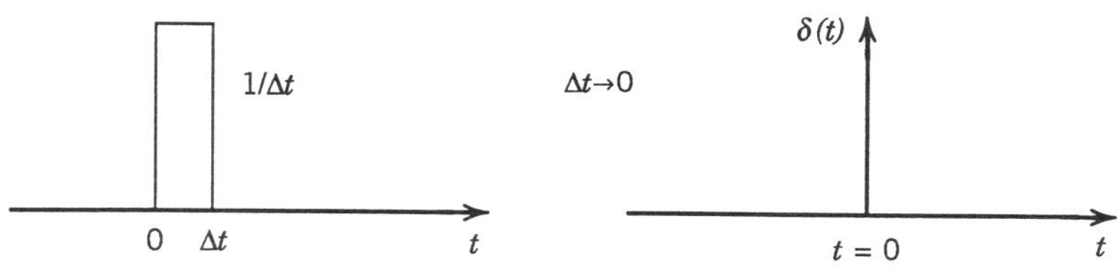

Fig. 3.2 Delta function

Time discretization represents an approximation of the continuous-time system where $\Delta t \neq 0$ replaces dt in the derivative, δ_{kj} replaces $\delta(t - \tau)$ in the covariance matrices of the stochastic processes, and $k \, \Delta t = t$:

$$[x(k+1) - x(k)]/\Delta t = F(t)x(k) + G(t)w(k). \tag{3.16}$$

This may be written, in a discrete-time form, as

$$\begin{aligned} x(k+1) &= F(k)x(k) + G(k)w(k), \\ z(k) &= H(k)x(k) + v(k), \end{aligned} \tag{3.17}$$

and

$$\begin{aligned} E\{w(k)w(j)^T\} &= Q(k)\,\delta_{kj} = [Q(t)/\Delta t]\,\delta_{kj} \\ E\{v(k)v(j)^T\} &= R(k)\,\delta_{kj} = [R(t)/\Delta t]\,\delta_{kj}, \end{aligned}$$

where the following substitutions were made:

$$\begin{aligned} F(k) &= I + F(t)\,\Delta t, \\ G(k) &= G(t)\,\Delta t, \\ H(k) &= H(t), \\ Q(k) &= Q(t)/\Delta t, \\ R(k) &= R(t)/\Delta t. \end{aligned} \tag{3.18}$$

For the system of Eq. (3.17), with substitutions of (3.18), the discrete Kalman filter is obtained from Eq. (3.9) as follows:

$$\begin{aligned} \hat{x}(k/k) &= F(k-1)\hat{x}(k-1/k-1) \\ &\quad + K(k)[z(k) - H(k)F(k-1)\hat{x}(k-1/k-1)], \quad (3.19) \\ K(k) &= \tilde{P}(k/k-1)H(k)^T\left[H(k)\tilde{P}(k/k-1)H(k)^T + R(k)\right]^{-1}, \\ \tilde{P}(k/k) &= [I - K(k)H(k)]\tilde{P}(k/k-1), \\ \tilde{P}(k/k-1) &= F(k-1)\tilde{P}(k-1/k-1)F(k)^T \\ &\quad + G(k-1)Q(k-1)G(k-1)^T, \\ \hat{x}(0/0) &= E\{x(0)\}, \\ \tilde{P}(0/0) &= E\left\{(\hat{x}(0/0) - x_0)(\hat{x}(0/0) - x_0)^T\right\} = P_0. \end{aligned} \tag{3.19}$$

In order to convert the discrete-time KF to the continuous time by taking $\Delta t \to 0$, first replace the continuous values of Eq. (3.18) into Eq. (3.19):

$$\begin{aligned}\hat{x}(k/k) &= [I + F(t)\,\Delta t]\hat{x}(k-1/k-1)\\ &\quad + K(k)/\Delta t[z(k) - H(k)(I + F(t)\,\Delta t)\hat{x}(k-1/k-1)]\,\Delta t.\end{aligned}$$

Reconstructing the differential equation in the form of Eq. (3.16), and defining $o(\Delta t)$ as the expression containing higher-order terms in Δt, for example, $\lim_{\Delta t \to 0} o(\Delta t) = 0$;

$$\begin{aligned}&[\hat{x}(k/k) - \hat{x}(k-1/k-1)]/\Delta t\\ &\quad = F(t)\hat{x}(k-1/k-1) + K(k)/\Delta t\\ &\qquad \times[z(k) - H(k)\hat{x}(k-1/k-1)] + o(\Delta t),\end{aligned}$$

$$K(k) = \tilde{P}(k/k-1)H(k)^T\left[H(k)\tilde{P}(k/k-1)H(k)^T + R(k)/\Delta t\right]^{-1},$$

$$K(k)/\Delta t = \tilde{P}(k/k-1)H(k)^T\left[H(k)\tilde{P}(k/k-1)H(k)^T\,\Delta t + R(k)\right]^{-1}.$$

Taking the limit as $\Delta t \to 0$, and $(k-1)\,\Delta t = t \to k\,\Delta t$, the KBF equations become

$$\begin{aligned}d\hat{x}(t/t)/dt &= F(t)\hat{x}(t/t) + K(t)[z(t) - H(t)\hat{x}(t/t)],\\ K(t) &= \lim_{\Delta t \to 0} K(k)/\Delta t = \tilde{P}(k/k-1)H(t)^T R(t)^{-1}\\ &= \tilde{P}(t/t)H(t)^T R(t)^{-1},\\ \tilde{P}(t/t) &= \lim_{\Delta t \to 0} \tilde{P}(k/k) = \lim_{\Delta t \to 0}[I - [K(k)/\Delta t]H(k)\,\Delta t]\tilde{P}(k/k-1)\\ &= \lim_{\Delta t \to 0} \tilde{P}(k/k-1),\end{aligned}$$

$$\begin{aligned}&\tilde{P}(k+1/)\\ &\quad = [I + F(t)\,\Delta t][I - K(k)H(k)]\tilde{P}(k/k-1)\\ &\qquad \times[I + F(t)\,\Delta t]^T + G(k)Q(t)G(k)^T\,\Delta t\\ &\quad = [I + F(t)\,\Delta t]\tilde{P}(k/k)[I + F(t)\,\Delta t]^T + G(k)Q(t)G(k)^T\,\Delta t\\ &\quad = \tilde{P}(k/k-1) + F(t)\tilde{P}(k/k-1)\,\Delta t\\ &\qquad + \tilde{P}(k/k-1)F(t)^T\,\Delta t - K(k)H(k)\tilde{P}(k/k-1)\\ &\qquad - F(t)K(k)H(k)\tilde{P}(k/k-1)\,\Delta t - K(k)H(k)\tilde{P}(k/k-1)F(t)^T\,\Delta t\\ &\qquad + F(t)\tilde{P}(k/k-1)F(t)^T\,\Delta t^2\\ &\qquad - F(t)K(k)H(k)\tilde{P}(k/k-1)F(t)^T\,\Delta t^2 + G(k)Q(t)G(k)^T\,\Delta t.\end{aligned} \tag{3.20}$$

Rearranging, substituting the continuous parameters, and dividing by Δt,

$$\begin{aligned}
&\left[\tilde{P}(k+1/k) - \tilde{P}(k/k-1)\right]/\Delta t \\
&\quad = F(t)\tilde{P}(k/k-1) + \tilde{P}(k/k-1)F(t)^T - K(t)H(t)\tilde{P}(k/k-1) \\
&\qquad + G(t)Q(t)G(t)^T + O(\Delta t),
\end{aligned} \tag{3.21}$$

where $K(k) = K(t)\,\Delta t \to 0$ and $O(\Delta t) \to 0$, as $\Delta t \to 0$,

$$\begin{aligned}
d\tilde{P}(t/t)/dt &= \lim_{\Delta t \to 0} \left[\tilde{P}(k+1/k) - \tilde{P}(k/k-1)\right]/\Delta t \\
&= F(t)\tilde{P}(t/t) + \tilde{P}(t/t)F(t)^T \\
&\quad - K(t)H(t)\tilde{P}(t/t) + G(t)Q(t)G(t)^T.
\end{aligned} \tag{3.22}$$

Equation (3.22) is known as the *Matrix Riccati equation*, because of its resemblance to the scalar quadratic Riccati equation, which has a known solution.

Collecting Eqs. (3.20) and (3.22), the *continuous-time Kalman–Bucy filter* (Figure 3.3) is obtained:

$$\begin{aligned}
d\hat{x}(t/t)/dt &= F(t)\hat{x}(t/t) + K(t)[z(t) - H(t)\hat{x}(t/t)], \\
K(t) &= \tilde{P}(t/t)H(t)^T R(t)^{-1}, \\
d\tilde{P}(t/t)/dt &= F(t)\tilde{P}(t/t) + \tilde{P}(t/t)F(t)^T \\
&\quad - K(t)H(t)\tilde{P}(t/t) + G(t)Q(t)G(t)^T.
\end{aligned} \tag{3.23}$$

If one wishes to obtain estimates of smoothing or prediction at some other time $T \neq t$, using the state transition matrix $\Phi(\cdot,\cdot)$ of system (3.12) obtains

$$\hat{x}(T/t) = \Phi(T/t)\hat{x}(t/t). \tag{3.24}$$

AN ILLUSTRATIVE EXAMPLE An illustrative example will demonstrate the method. Consider, as in the discrete-time case, a scalar system with no process noise:

$$\begin{aligned}
dx(t)/dt &= 0, \qquad x(0), \\
z(t) &= x(t) + v(t),
\end{aligned}$$

where $x(0)$ is a random variable with x_0 and p_0, $v(t) \sim N(0, r_0)$, $\forall\, t \in [0, \infty]$, $f(t) = g(t) = q(t) = 0$, $h(t) = 1$; the KBF yields a scalar Riccati equation,

$$d\tilde{P}/dt = -\tilde{P}(t/t)^2/r_0, \tag{3.25}$$

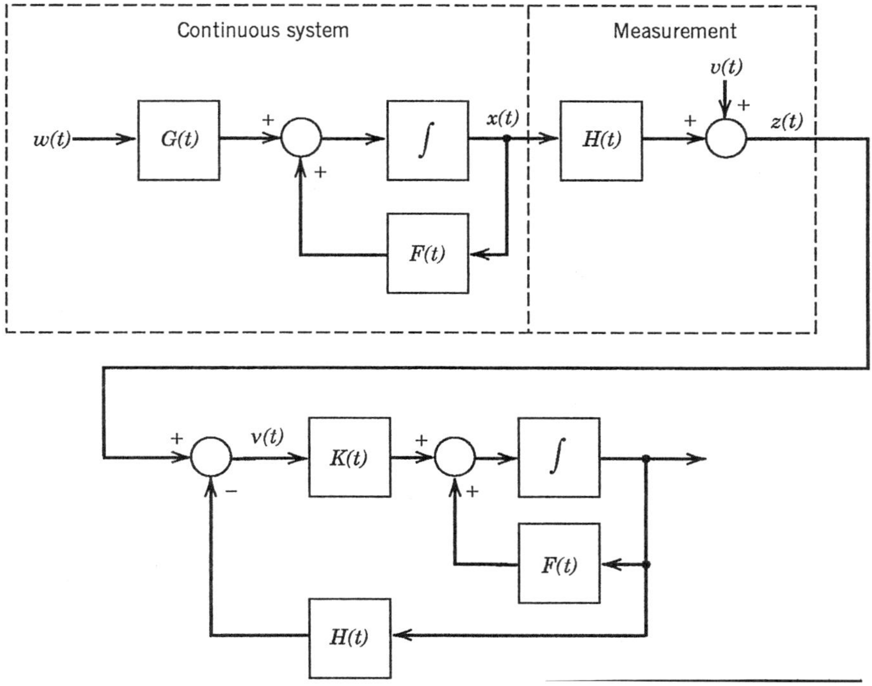

Fig. 3.3 System model and continuous-time Kalman–Bucy filter

which may be linearized by substituting $\tilde{P}(t/t) = 1/y$ and $d\tilde{P}/dt = -1/y^2$, $p_0 = 1/y_0$ in (3.25),

$$-\dot{y}/y^2 = -1/y^2 r_0 \Rightarrow \dot{y} = 1/r_0, \qquad y_0. \tag{3.26}$$

Solving Eq. (3.26),

$$y(t) - y_0 = 1/p(t) - 1/p_0 = t/r_0$$

or

$$\tilde{P}(t/t) = p_0 r_0/(r_0 + t).$$

The KBF estimator is given by

$$\begin{aligned}\hat{x}(t/t) &= K(t)[z(t) - \hat{x}(t/t)] \\ &= \tilde{P}(t/t)/r_0[z(t) - \hat{x}(t/t)] = p_0 r_0/(r_0 + t)[z(t) - \hat{x}(t/t),\end{aligned}$$

where the KBF gain is

$$K(t) = \tilde{P}(t/t)/r_0 = p_0/(r_0 + t).$$

The above derivations provide a full description of the optimal filters and predictors used to define the future behavior of a linear dynamic system corrupted with Gaussian white noise. If the noise is not Gaussian they still represent optimal filters and predictors in the minimum mean-square sense. The only case not discussed in this chapter is the Smoothing problem, which is not required for the treatment of the material in the following. An excellent presentation of this material is given in Meditch (1969).

3.5 SOLUTION OF THE MATRIX RICCATI EQUATION

The Matrix Riccati equation, which defines the propagation of the error covariance, is quadratic in $\tilde{P}(t/t)$, independent of the measurements, resembling the scalar Riccati equation, solved in the previous example:

$$\begin{aligned} d\tilde{P}/dt &= F(t)\tilde{P}(t/t) + \tilde{P}(t/t)F(t)^T + G(t)Q(t)G(t)^T \\ &\quad - \tilde{P}(t/t)H(t)^T R(t)^{-1} H(t)\tilde{P}(t/t), \\ \tilde{P}(0/0) &= P_0. \end{aligned} \tag{3.27}$$

This equation could be solved in a manner similar to the scalar case, by replacing $\tilde{P}(t/t) = S(t)^{-1} - A$, and solving its homogeneous part first:

$$\begin{aligned} dS/dt &= S(t)\left[F(t) + AH^T R^{-1} H\right] \\ &\quad + \left[F(t) + AH^T R^{-1} H\right]^T S(t) - H(t)^T R(t)^{-1} H(t), \end{aligned} \tag{3.28}$$

and then solve the forced equation (3.28), using the method of variation of parameters, $S(t) = \Psi(t)C(t)$, where $C(t)$ is the initial parameter, $\Psi(t)$ is the state transition matrix of (3.28), and the matrix A is given by $A = A^T > 0 \ni FA + AF^T + AH^T R^{-1} HA$, such that

$$\begin{aligned} S(t) &= \Psi(t)S(0) - \int_0^t \Psi(t-\tau)H^T R^{-1} H\,d\tau, \qquad S(0) = (P_0 + A)^{-1}, \\ \tilde{P}(t/t) &= S(t)^{-1} - A. \end{aligned} \tag{3.29}$$

The solution obtained in this manner, even though quite elegant, presents considerable computational difficulties. Therefore, a solution through linearization by augmenting the dimension of the equation and solving a system of $2n$ differential equations, resembling the states and costates of the optimal control problem, is preferred.

Let $\lambda(t) = \tilde{P}(t/t)y(t)$, and $d\lambda/dt = d\tilde{P}/dt\, y + \tilde{P}\, dy/dt$, where the arguments of the variables were dropped for simplicity of notation, without any loss of generality. Assuming also that

$$dy/dt = -F^T y + H^T R^{-1} H \tilde{P} y, \tag{3.30}$$

then λ satisfies

$$d\lambda/dt = F\tilde{P}y + \tilde{P}F^T y + GQG^T y - \tilde{P}H^T R^{-1} H \tilde{P} y - \tilde{P}F^T y + \tilde{P}H^T R^{-1} H \tilde{P} y. \tag{3.31}$$

Simplifying Eqs. (3.30) and (3.31), by observing that $\tilde{P}y = \lambda$, they become the equations to be solved simultaneously to obtain $\tilde{P}(t/t)$:

$$\begin{aligned} dy/dt &= -F^T y + H^T R^{-1} H \lambda, \\ d\lambda/dt &= F\lambda + GQG^T y. \end{aligned} \tag{3.32}$$

In a matrix vector form Eqs. (3.32) become:

$$\begin{bmatrix} dy/dt \\ d\lambda/dt \end{bmatrix} = \begin{bmatrix} -F^T & H^T R^{-1} H \\ GQG^T & F \end{bmatrix} \begin{bmatrix} y \\ \lambda \end{bmatrix}. \tag{3.33}$$

Solving (3.33) simultaneously, and considering Φ_{yy}, $\Phi_{\lambda y}$, $\Phi_{y\lambda}$, and $\Phi_{\lambda\lambda}$ the appropriate partitions of the state transition matrix are

$$\begin{bmatrix} y(t) \\ \lambda(t) \end{bmatrix} = \begin{bmatrix} \Phi_{yy}(t) & \Phi_{y\lambda}(t) \\ \Phi_{\lambda y}(t) & \Phi_{\lambda\lambda}(t) \end{bmatrix} \begin{bmatrix} y(0) \\ \lambda(0) \end{bmatrix}. \tag{3.34}$$

Considering the relation $\lambda(t) = \tilde{P}(t/t)y(t)$, and on the initial conditions $\lambda(0) = P_0 y(0)$, and removing $y(0)$ from both equations, one obtains the solution of the Matrix Riccati equation:

$$\tilde{P}(t/t) = \left[\Phi_{\lambda y}(t) + \Phi_{\lambda\lambda}(t)P_0\right]\left[\Phi_{yy}(t) + \Phi_{y\lambda}(t)P_0\right]^{-1}. \tag{3.35}$$

This solution may be easily obtained by using any software package designed to solve a linear time-varying system.

This completes the derivation of the continuous-time Kalman–Bucy filter.

EXAMPLE Consider the system of Figure 2.3, of deterministic least-square parameter estimation, now used for state estimation:

$$\begin{bmatrix} dx_1/dt \\ dx_2/dt \end{bmatrix} = \begin{bmatrix} 0 & 1 \\ -\omega_n^2 & -2\zeta\omega_n \end{bmatrix} \begin{bmatrix} x_1 \\ x_2 \end{bmatrix} + \begin{bmatrix} 0 \\ 1 \end{bmatrix} w(t),$$

$$z = x_2 + v(t),$$

where $x(0) \sim N(x_0, P_0)$ is the initial vector, and $w(t) \sim N(0, q^2)$, $v(t) \sim N(0, r^2)$ are the process and measurement noise, respectively. The Kalman–Bucy estimator is, from Eq. (3.23):

$$\begin{bmatrix} dx_1/dt \\ dx_2/dt \end{bmatrix} = \begin{bmatrix} 0 & 1 \\ -\omega_n^2 & -2\zeta\omega_n \end{bmatrix} \begin{bmatrix} x_1 \\ x_2 \end{bmatrix} + \frac{1}{r^2} \begin{bmatrix} p_2 \\ p_3 \end{bmatrix} [z - x_2].$$

The Matrix Riccati equation for $\tilde{P}(0) = P_0$ is given in a 2×2 matrix form by

$$d\tilde{P}/dt = F\tilde{P} + \tilde{P}F^T + \begin{bmatrix} 0 \\ 1 \end{bmatrix} [0 \quad 1] q^2 - \frac{1}{r^2} \tilde{P} \begin{bmatrix} 0 \\ 1 \end{bmatrix} [0 \quad 1] \tilde{P}$$

or

$$\dot{p}_{11} = 2p_{11} - 1/r^2 p_{22}^2,$$

$$\dot{p}_{12} = -\omega_n^2 p_{11} - 2\zeta\omega_n p_{12} + p_{22} - 1/r^2 p_{12} p_{22},$$

$$\dot{p}_{22} = -2\omega_n^2 p_{12} - 2\zeta\omega_n p_{22} + q^2 - 1/r^2 p_{22}.$$

The solution of this equation requires a complex computer program as in Lewis (1986). The steady-state analytic solution is given by Lewis:

$$p_{11}(\infty) = 2\zeta r^2/\omega_n \left(\sqrt{1 + q^2/4\zeta^2\omega_n^2 r^2} - 1\right),$$

$$p_{12}(\infty) = 0$$

$$p_{22}(\infty) = 2\zeta r^2 \omega_n \left(\sqrt{1 + q^2/4\zeta^2\omega_n^2 r^2} - 1\right).$$

The solution using the linearization procedure by doubling the number of equations is given by assuming the variables $y(t)$ and $\lambda(t)$ related by

$$\lambda(t) = \tilde{P}(t/t) y(t)$$

and solving the set of $2n$ homogeneous differential equations:

$$\begin{bmatrix} dy/dt \\ d\lambda/dt \end{bmatrix} = \begin{bmatrix} -F^T & H^T R^{-1} H \\ GQG^T & F \end{bmatrix} \begin{bmatrix} y(t) \\ \lambda(t) \end{bmatrix}$$

$$F = \begin{bmatrix} 0 & 1 \\ -\omega_n^2 & -2\zeta\omega_n \end{bmatrix} H^T R^{-1} H = \begin{bmatrix} 0 & 0 \\ 0 & r^{-2} \end{bmatrix} GQG^T = \begin{bmatrix} 0 & 0 \\ 0 & q^2 \end{bmatrix}.$$

The solution of this set of equations is given by its state transition matrix

$$\Phi(t) = \begin{bmatrix} \Phi_{yy} & \Phi_{y\lambda} \\ \Phi_{y\lambda} & \Phi_{\lambda\lambda} \end{bmatrix},$$

where for

$$s = \sqrt{a + \sqrt{a^2 + \omega_n^2}}, \qquad v = \sqrt{a - \sqrt{a^2 + \omega_n^2}},$$

and

$$a = q^2/2r^2 - \omega_n^2(1 - 2\zeta^2),$$

$$\Phi_{yy} = \begin{bmatrix} A_1 e^{st} + A_2 e^{-st} + A_3 e^{vt} + A_4 e^{-vt} & B_1 e^{st} + B_2 e^{-st} + B_3 e^{vt} + B_4 e^{-vt} \\ C_1 e^{st} + C_2 e^{-st} + C_3 e^{vt} + C_4 e^{-vt} & D_1 e^{st} + D_2 e^{-st} + D_3 e^{vt} + D_4 e^{-vt} \end{bmatrix},$$

$$\Phi_{y\lambda} = \begin{bmatrix} E_1 e^{st} + E_2 e^{-st} + E_3 e^{vt} + E_4 e^{-vt} & F_1 e^{st} + F_2 e^{-st} + F_3 e^{vt} + F_4 e^{-vt} \\ G_1 e^{st} + G_2 e^{-st} + G_3 e^{vt} + G_4 e^{-vt} & H_1 e^{st} + H_2 e^{-st} + H_3 e^{vt} + H_4 e^{-vt} \end{bmatrix},$$

$$\Phi_{\lambda\lambda} = \begin{bmatrix} K_1 e^{st} + K_2 e^{-st} + K_3 e^{vt} + K_4 e^{-vt} & L_1 e^{st} + L_2 e^{-st} + L_3 e^{vt} + L_4 e^{-vt} \\ M_1 e^{st} + M_2 e^{-st} + M_3 e^{vt} + M_4 e^{-vt} & N_1 e^{st} + N_2 e^{-st} + N_3 e^{vt} + N_4 e^{-vt} \end{bmatrix},$$

and the solution of the Matrix Riccati equation is given by Eq. (3.35):

$$\tilde{P}(t/t) = \left[\Phi_{\lambda y}(t) + \Phi_{\lambda\lambda}(t) P_0\right]\left[\Phi_{yy}(t) + \Phi_{y\lambda}(t) P_0\right]^{-1},$$

where the coefficients are selected to satisfy the state transition matrix.

3.6 CONTROLLABILITY, OBSERVABILITY, STABILITY

Optimality of the estimator does not guarantee its stability, as in the discrete-time case. Additional conditions must be imposed on the system.

3.6.1 Stochastic Controllability

In the absence of measurements and complete a priori information and constant coefficients F, G, H, Q, R, the continuous-time Matrix Riccati equation is identical to the covariance propagation equation:

$$d\tilde{P}/dt = F\tilde{P} + \tilde{P}F^T + GQG^T, \qquad \tilde{P}(0/0) = 0. \tag{3.36}$$

Because Eq. (36) is linear in $\tilde{P}$, its solution is given by

$$\tilde{P}(t/t) = \int_0^t \Phi(t,\tau)GQG^T\Phi(t,\tau)^T\,d\tau, \tag{3.37}$$

where $\Phi(t,\tau)$ is the state transition matrix of system (3.12). If the integral of (3.37) is positive definite and bounded for any $t > 0$, the system is said to be *uniformly completely controllable.*

3.6.2 Stochastic Observability

In the absence of process noise and a priori information the continuous-time Matrix Riccati equation becomes,

$$d\tilde{P}/dt = F\tilde{P} + \tilde{P}F^T - \tilde{P}H^TR^{-1}H\tilde{P}, \qquad \tilde{P}(0/0) \to \infty. \tag{3.38}$$

Using the matrix identity, $d\tilde{P}^{-1}/dt = -\tilde{P}^{-1}\,d\tilde{P}/dt\tilde{P}^{-1}$, Eq. (3.38) is rewritten as

$$d\tilde{P}^{-1}/dt = -\tilde{P}^{-1}F - F^T\tilde{P}^{-1} + H^TR^{-1}H, \qquad \tilde{P}(0/0)^{-1} = 0. \tag{3.39}$$

The solution of (3.39) is

$$\tilde{P}(t/t)^{-1} = \int_0^t \Phi^T(\tau,t)H^TR^{-1}H\Phi(\tau,t)\,d\tau, \tag{3.40}$$

where $\Phi(t,\tau)$ is the system transition matrix. If the integral is positive definite for some $t > 0$, then $\tilde{P}(t/t)^{-1} > 0$, and $0 < \tilde{P}(t/t) < \infty$; for example, it is possible to obtain information about the states that were initially unknown. System (3.12) is said to be *uniformly completely observable.*

3.6.3 Asymptotic Stability

Asymptotic stability of the equilibrium of system (3.12) deals with the unforced part of the estimator equation:

$$d\hat{x}(t/t)/dt = [F - K(t)H]\hat{x}(t/t), \tag{3.41}$$

when $\hat{x}(t/t) \to 0$, as $t \to \infty$, for any initial condition $\hat{x}(0/0)$. However, if the system (3.12) is *stochastically uniformly completely controllable and observable*, then it is also *asymptotically stable* and the covariance matrix $\tilde{P}(t/t)$ is *unique*.

The covariance propagation equation (3.35), for a given $\tilde{P}(0/0) = P_0$, has a solution easily obtainable from Eq. (3.35), by setting $\Phi_{yy} = (e^{-Ft})^T$, $\Phi_{\lambda\lambda} = e^{-Ft}$, and $\Phi_{y\lambda} = 0$,

$$\tilde{P}(t/t) = \Phi_{\lambda y}(t)\Phi_{\lambda\lambda}(t)^T + \Phi_{\lambda\lambda}(t)P_0\Phi_{\lambda\lambda}(t)^T. \tag{3.42}$$

Equation (3.42) in the limit becomes $P_\infty = \lim_{t\to\infty} \Phi_{\lambda y}(t)\Phi_{\lambda\lambda}(t)^T$, independent of P_0.

3.7 STEADY-STATE FILTER: THE WIENER FILTER

It has been shown that the Kalman–Bucy filter and the Matrix Riccati equation exist, and are stable as $t \to \infty$, when the system coefficients F, G, H, and noise covariances are constant. Let $\tilde{P}(\infty/\infty) = P_\infty$. Then the Kalman–Bucy filter becomes

$$d\hat{x}(t/t) = F\hat{x}(t/t) + K_\infty[z(t) - H\hat{x}(t/t)], \qquad \hat{x}(0/0) = x_0,$$
$$K_\infty = P_\infty H^T R^{-1}, \tag{3.43}$$

and the Matrix Riccati equation becomes algebraic:

$$FP_\infty + P_\infty F^T + GQG^T - P_\infty H^T R^{-1} H P_\infty = 0, \tag{3.44}$$

and its solution becomes

$$P_\infty = \lim_{\tau\to 0}\left[\Phi_{\lambda y}(\tau) + \Phi_{\lambda\lambda}(\tau)P_0\right]\left[\Phi_{yy} + \Phi_{y\lambda}P_0\right]^{-1}. \tag{3.45}$$

Taking the Laplace transform of Eq. (3.36), one obtains the *Wiener filter*:

$$d\hat{x}(t/t)/dt = (F - K_\infty H)\hat{x}(t/t) + K_\infty z(t) \to$$
$$(sI - F + K_\infty H)\hat{X}(s) = K_\infty Z(s),$$
$$\hat{X}(s) = (sI - F + K_\infty H)^{-1}K_\infty Z(s). \tag{3.46}$$

An excellent reference for the above material is given in Gelb (1974).

3.8 NONLINEAR STATE ESTIMATION

State estimation of nonlinear discrete- or continuous-time dynamic systems present considerable difficulties, even with the assumption of Gaussian initial conditions, process, and measurement additive noises. Gaussian noise preserves its Gaussian property when it propagates through a linear system, thus yielding a most-probable as well as an MMSE estimator. This is not true in the case of a nonlinear system, where at best, an MMSE estimator should be expected. Further complications arise with the fact that the expected value of a nonlinear function is not equal to the nonlinear function of the expected value of the states; for example,

$$E\{f(x(t))\} \neq f(E\{x(t)\}), \tag{3.47}$$

requiring the knowledge of the density function $p(x(t))$ at every $t < \infty$. Therefore, in the case of nonlinear estimation, one should be satisfied with some kind of approximation. Additional problems arise in the case of continuous-time estimation. The Matrix Riccati equation is no longer independent of the measurements, due to integration problems of the Gaussian stochastic processes. Derivation of the best known approximate MMSE estimates of the discrete-time, continuous-time, and continuous-discrete-time, the extended and linearized filters, are given in the following.

3.9 DISCRETE-TIME EXTENDED KALMAN FILTER

Consider the nonlinear discrete-time system, with properties similar to Eqs. (3.1) and (3.2):

$$\begin{aligned} x(k+1) &= f(x(k)) + G(k)w(k), \qquad x(0), \\ z(k) &= h(x(k)) + v(k), \end{aligned} \tag{3.48}$$

for a measurement m vector $z(k)$, state n vector $x(k)$, and nonlinear twice-differentiable functions $f(\cdot)$, $h(\cdot)$. The stochastic variables are $x(0)$, the n-dimensional initial state; $w(k)$, an r-dimensional process noise; and $v(k)$, an m-dimensional measurement noise, which for any time k have the following properties:

$$\begin{aligned} &E\{x(0)\} = x_0, \quad E\{(x(0) - x_0)(x(0) - x_0)^T\} = P_0, \\ &E\{w(k)\} = 0, \quad E\{w(k)w(j)^T\} = Q(k)\,\delta_{kj}, \quad E\{v(k)v(j)^T\} = R(k)\,\delta_{kj}, \\ &E\{v(k)\} = 0, \quad E\{w(k)x(0)^T\} = E\{v(k)x(0)^T\} = E\{w(k)v(j)^T\} = 0. \end{aligned} \tag{3.49}$$

Assuming that $\hat{x}(k/k)$ is the MMSE estimate of the state, one may expand the nonlinear terms in Taylor series about $\hat{x}(k/k)$:

$$
\begin{aligned}
f(x(k)) &= f(\hat{x}(k/m)) + F(\hat{x}(k/m))[x(k) - \hat{x}(k/m)] \\
&\quad + o(x(k) - \hat{x}(k/m)), \\
h(x(k)) &= h(\hat{x}(k/m)) + H(\hat{x}(k/m))[x(k) - \hat{x}(k/m)] \\
&\quad + o(x(k) - \hat{x}(k/m)),
\end{aligned} \tag{3.50}
$$

where F and H represent the Jacobian matrices, with respect to x;

$$
\begin{aligned}
F(x,k) &= \partial f(x,k)/\partial x, \\
H(x,k) &= \partial h(x,k)/\partial x.
\end{aligned} \tag{3.51}
$$

A first-order approximation of the conditional expectation of the nonlinear terms is obtained as

$$
\begin{aligned}
\hat{f}(k) &= E\{f(x(k))/Z^k\} \approx f(\hat{x}(k/k)), \\
\hat{h}(k) &= E\{h(x(k))/Z^k\} \approx h(\hat{x}(k/k)).
\end{aligned} \tag{3.52}
$$

Then, for some filter gain $K(k)$, the state estimate update and state estimator are, respectively,

$$
\begin{aligned}
\hat{x}(k/k-1) &= f(\hat{x}(k-1/k-1)), \\
\hat{x}(k/k) &= \hat{x}(k/k-1) + K(k)[z(k) - h(\hat{x}(k/k-1))].
\end{aligned} \tag{3.53}
$$

The error covariance matrix $\tilde{P}(k/k)$ is derived by the following arguments:

$$
\begin{aligned}
\tilde{x}(k/k) &= \hat{x}(k/k) - x(k) = \tilde{x}(k/k-1) \\
&\quad + K(k)[h(x(k)) - h(\hat{x}(k/k-1))] \\
&\quad + K(k)v(k) \approx \tilde{x}(k/k-1) \\
&\quad - K(k)H(\hat{x}(k/k-1))\tilde{x}(k/k-1) + K(k)v(k).
\end{aligned} \tag{3.54}
$$

Then

$$\begin{aligned}\tilde{P}(k/k) &= E\{\tilde{x}(k/k)\tilde{x}(k/k)^T\}\\ &\approx \tilde{P}(k/k-1) - K(k)H(x(k/k-1))\tilde{P}(k/k-1)\\ &\quad \times H(\hat{x}(k/k-1))^T K(k) + \tilde{P}(k/k-1)H(\hat{x}(k/k-1))^T K(k)^T\\ &\quad - K(k)H(\hat{x}(k/k-1))\tilde{P}(k/k-1) + K(k)R(k)K(k)^T,\end{aligned}$$

$$\begin{aligned}\tilde{P}(k/k-1) &= E\{\tilde{x}(k/k-1)\tilde{x}(k/k-1)^T\}\\ &= E\{[\hat{x}(k/k-1) - x(k)][\hat{x}(k/k-1) - x(k)]^T\}\\ &= E\{[f(\hat{x}(k/k-1)) - f(x(k) - Gw(k)]\\ &\qquad \times [f(\hat{x}(k/k-1)) - f(x(k) - Gw(k)]^T\}\\ &= F(\hat{x}(k/k-1))\tilde{P}(k-1/k-1)F(\hat{x}(k/k-1))^T\\ &\quad + G(k-1)Q(k-1)G(k-1)^T. \end{aligned} \tag{3.55}$$

When a minimum variance estimate is sought, a $K(k)$ should be selected that minimizes the error variance:

$$\sigma_{\tilde{x}}^2 = E\{\tilde{x}(k/k)^T\tilde{x}(k/k)\} = \operatorname{Trace}[\tilde{P}(k/k)]. \tag{3.56}$$

The minimization procedure, for example, $\partial\sigma_{\tilde{x}}^2/\partial K(k)$, yields

$$\begin{aligned}K(k) &= \tilde{P}(k/k-1)H(\hat{x}(k/k-1))^T\\ &\quad \times \left[H(\hat{x}(k/k-1))\tilde{P}(k/k-1)H(\hat{x}(k/k-1))^T + R(k)\right]^{-1}.\end{aligned} \tag{3.57}$$

Substituting back to Eq. (3.55), and after some algebra,

$$\tilde{P}(k/k) = [I - K(k)H(\hat{x}(k/k-1))]\tilde{P}(k/k-1). \tag{3.58}$$

TABLE 3.3 Summary of Discrete-Time Extended Kalman Filter

System model	$x(k+1) = f(k, x(k)) + G(k)w(k), \qquad w(k) \sim N(0, Q(k))$
Measurement model	$z(k) = H(k)x(k) + v(k), \qquad v(k) \sim N(0, R(k))$
Initial conditions	$E\{x(0)\} = \hat{x}_0, \qquad E\{[x(0) - \hat{x}_0][x(0) - \hat{x}_0]^T\} = P_0$
Other assumptions	$R(k)^{-1} > 0$ exists, $\qquad E\{w(k)v(j)^T\} = 0$
State estimate	$\hat{x}(k/k) = \hat{x}(k/k-1) + K(k)[z(k) - h(\hat{x}(k/k-1))]$,
	$\hat{x}(k/k-1) = f(\hat{x}(k-1/k-1)); \qquad \hat{x}(0) = \hat{x}_0$
Error covariance	$\tilde{P}(k/k) = [I - K(k)H(\hat{x}(k/k-1))]\tilde{P}(k/k-1)$;
	$\tilde{P}(k/k-1) = F(\hat{x}(k/k-1))\tilde{P}(k-1/k-1)F(\hat{x}(k/k-1))^T$
	$+G(k-1)Q(k-1)G(k-1)^T, \qquad \tilde{P}(0/0) = \tilde{P}_0$
KB gain matrix	$K(k) = \tilde{P}(k/k-1)H(\hat{x}(k/k-1))^T$
	$\times[H(\hat{x}(k/k-1))\tilde{P}(k/k-1)H(\hat{x}(k/k-1))^T + R(k)]^{-1}$
Definitions	$F(k, \hat{x}(k)) = \partial f(k, x(k))/\partial x \vert x(k) = \hat{x}(k)$
	$H(k, \hat{x}(k)) = \partial h(k, x(k))/\partial x \vert x(k) = \hat{x}(k)$

Collecting Eqs. (3.53), (3.57), and (3.58), one obtains the *discrete-time extended Kalman filter* (see Table 3.3)

$$
\begin{aligned}
\hat{x}(k/k-1) &= f(\hat{x}(k-1/k-1)), \\
\hat{x}(k/k) &= \hat{x}(k/k-1) + K(k)[z(k) - h(\hat{x}(k/k-1))], \\
K(k) &= \tilde{P}(k/k-1)H(\hat{x}(k/k-1))^T \\
&\quad \times\left[H(\hat{x}(k/k-1))\tilde{P}(k/k-1)H(\hat{x}(k/k-1))^T + R(k)\right]^{-1}, \\
\tilde{P}(k/k) &= [I - K(k)H(\hat{x}(k/k-1))]\tilde{P}(k/k-1), \\
\tilde{P}(k/k-1) &= F(\hat{x}(k/k-1))\tilde{P}(k-1/k-1)F(\hat{x}(k/k-1))^T \\
&\quad + G(k-1)Q(k-1)G(k-1)^T.
\end{aligned}
\tag{3.59}
$$

The extended Kalman filter gives very good estimates of the states of a discrete-time system if their trajectories are smooth enough. This is the reason it found excellent applications in estimating trajectories of space vehicles. Divergence of the filter is observed when the estimates and their initial conditions are not reasonably close to the true values.

3.10 DISCRETE-TIME LINEARIZED KALMAN FILTER

A discrete-time Kalman filter may be obtained, if system (3.48) is linearized before processing the measurement data, about a vector $x'(t)$, say, a nominal solution of the system

$$x'(k+1) = f(x'(k)), \qquad x'(0) = x_0. \tag{3.60}$$

The resulting estimator looks exactly the same as the extended Kalman filter, by replacing $\hat{x}(k/k-1)$ with $x'(k)$ in the functions $f(\cdot)$, $h(\cdot)$, $F(\cdot)$, and $H(\cdot)$ of Eq. (3.59):

$$\begin{aligned}\hat{x}(k/k) &= f(x'(k-1)) + F(x'(k-1)) \\ &\quad \times [x(k-1/k-1) - x'(k) + K(k)[z(k) - h(x'(k)) \\ &\qquad - H(x'(k))F(x'(k-1))(x(k-1/k-1) - x'(k-1))], \\ K(k) &= \tilde{P}(k/k-1)H(x'(k))^T \\ &\quad \times \left[H(x'(k))\tilde{P}(k/k-1)H(x'(k))^T + R(k)\right]^{-1}, \\ \tilde{P}(k/k) &= [I - K(k)H(x'(k))]\tilde{P}(k/k-1), \\ \tilde{P}(k/k-1) &= F(x'(k))\tilde{P}(k-1/k-1)F(x'(k))^T \\ &\quad + G(k-1)Q(k-1)G(k-1)^T. \end{aligned} \tag{3.61}$$

The discrete-time linearized Kalman filter does not produce good estimates of the system (3.1) trajectories. Sometimes this filter diverges, when the nominal trajectory is not close to the true trajectory, or x_0 is not a good estimate of the initial conditions. The extended Kalman filter always gives better estimates than the linearized Kalman filter.

3.11 CONTINUOUS-TIME EXTENDED KALMAN FILTER

The nonlinear state estimation problem for continuous-time dynamic systems is further complicated by the fact that the integral of a Gaussian stochastic process does not exist in the Riemann sense. Instead, it may be assumed that the process and measurement noise are differentials of a Wiener–Levy process whose covariances are proportional to time, and therefore of the same order as the system variables. Of course, this difficulty arises only in the case of Gaussian random processes, which are the most popular in state estimation problems due to the central limit theorem.

In order to avoid the use of Itô calculus in the continuous-time domain, which requires mathematics beyond the scope of this text, an approach similar to the derivation of the linear continuous-time Kalman filter from its discrete-time version, for example, Eqs. (3.16)–(3.23), will be followed. Consider the following continuous-time stochastic dynamic system:

$$\begin{aligned} dx/dt &= f(x(t)) + G(t)w(t), \qquad x(0), \\ z(t) &= h(x(t)) + v(t), \end{aligned} \tag{3.62}$$

where $x(t)$ is the state of the system at time t, $z(t)$ is the current m-dimensional measurement vector, $f(x(t))$, $h(x(t))$ are twice-differentiable nonlinear functions of their arguments, and $G(t)$ is an $n \times r$ matrix of known coefficients. The stochastic variable $x(0)$ is the n-dimensional Gaussian initial state, $w(t)$ is the r-dimensional process noise, $v(t)$ is the m-dimensional measurement noise, defined for any $t \in [0, \infty]$, with the same properties as with the linear case, of Eqs. (3.13):

$$
\begin{aligned}
E\{x(0)\} &= x_0, \qquad E\{(x(0) - x_0)(x(0) - x_0)^T\} = P_0, \\
E\{w(t)\} &= 0, \qquad E\{w(t)w(\tau)^T\} = Q(t)\,\delta(t - \tau), \\
E\{v(t)\} &= 0, \qquad E\{v(t)v(\tau)^T\} = R(t)\,\delta(t - \tau).
\end{aligned}
\tag{3.63}
$$

The above properties indicate that the stochastic processes are independent identically distributed (i.i.d.), and possess finite means and covariances. In addition, Eq. (3.14),

$$
E\{x(0)w(t)^T\} = E\{x(0)v(t)^T\} = E\{w(t)v(\tau)^T\} = 0. \tag{3.64}
$$

System (3.62) is now discretized by considering a finite interval $\Delta t \neq 0$, and $t = k\,\Delta t$, to give the following discrete-time equations:

$$
\begin{aligned}
x(k+1) &= [x(k) + f(x(k))\,\Delta t] + G(t)w(k)\,\Delta t, \\
z(k) &= h(x(k)) + v(k), \\
Q(k) &= Q(t)/\Delta t; \qquad R(k) = R(t)/\Delta t.
\end{aligned}
\tag{3.65}
$$

Then, for some filter gain $K(k) = K(t)\,\Delta t$, $K(t) < \infty$, the state estimate update and state estimator are, respectively,

$$
\begin{aligned}
\hat{x}(k+1/k) &= \hat{x}(k/k) + \hat{f}(x(k))\,\Delta t, \\
\hat{x}(k+1/k+1) &= \hat{x}(k+1/k) + K(k+1)\big[z(k+1) - \hat{h}(x(k+1))\big].
\end{aligned}
\tag{3.66}
$$

Following the same linearization procedure as with the discrete-time extended Kalman filter, Eqs. (3.50)–(3.52), yield

$$
\begin{aligned}
f(x(k)) &= f(\hat{x}(k/m)) - F(\hat{x}(k/m))\tilde{x}(k/m) \\
&\quad + \tfrac{1}{2}\tilde{x}(k/m)^T\,\partial^2 f/\partial x^2\,\tilde{x}(k/m) + o\big(\tilde{x}(k/m)^2\big), \\
h(x(k)) &= h(\hat{x}(k/k)) - H(\hat{x}(k/k))\tilde{x}(k/k) \\
&\quad + \tfrac{1}{2}\tilde{x}(k/k)^T\,\partial^2 h/\partial x^2\,\tilde{x}(k/k) + o\big(\tilde{x}(k/k)^2\big),
\end{aligned}
\tag{3.67}
$$

where F and H represent the Jacobian matrices of f and h, respectively, and $\partial^2 h/\partial x^2$ is a tensor of the second partials of h with respect to x, needed for the derivation of $\tilde{P}(t/t)$:

$$F(x,k) = \partial f(x,k)/\partial x,$$
$$H(x,k) = \partial h(x,k)/\partial x,$$
$$\partial^2 \alpha/\partial x^2 = \left[\partial^2 \alpha_i/\partial x_j\, \partial x_k : i = 1 \cdots m;\ j,k = 1 \cdots n\right], \qquad \alpha = f,h.$$

From Eqs. (3.65), (3.66), and $K(k) = K(t)\,\Delta t$, the estimation error is given by

$$\begin{aligned}
&\hat{x}(k+1/k+1) \\
&\quad = \hat{x}(k+1)/k) + K(k+1)\left[z(k+1) - \hat{h}(x(k+1))\right] \\
&\quad = \hat{x}(k/k) + [f(\hat{x}(k/k)) - f(x(k))]\,\Delta t \\
&\qquad + K(t)\left[z(k+1) - \hat{h}(x(k+1))\right]\Delta t - G(k)w(k) \\
&\quad = \hat{x}(k/k) + \Delta f(\hat{x}(k/k))\,\Delta t + K(t)\left[z(k+1) - \hat{h}(x(k+1))\right]\Delta t \\
&\qquad - G(k)w(k)\,\Delta t.
\end{aligned} \tag{3.68}$$

A first-order approximation of the conditional expectation of the nonlinear terms is obtained as

$$\begin{aligned}
\hat{f}(x(k)) &= E\{f(x(k))/Z^k\} \approx f(\hat{x}(k/k)), \\
\hat{h}(x(k+1)) &= E\{h(x(k+1))/Z^k\} \approx h(\hat{x}(k+1/k)).
\end{aligned} \tag{3.69}$$

Substituting into Eq. (3.66) yields

$$\begin{aligned}
&[\hat{x}(k+1/k+1) - \hat{x}(k/k)]/\Delta t \\
&\qquad = f(\hat{x}(k/k)) + K(t)[z(k+1) - h(\hat{x}(k+1/k)].
\end{aligned}$$

Taking the limit as $\Delta t \to \infty$, and $k \to t$, the state estimator is obtained:

$$d\hat{x}(t/t)/dt = f(\hat{x}(t/t)) + K(t)[z(t) - h(\hat{x}(t/t))]. \tag{3.70}$$

The error covariance matrix $\tilde{P}(t/t)$ is derived from the discrete-time error;

$$\begin{aligned}
&\tilde{x}(k+1/k) \\
&\quad = \hat{x}(k+1/k) - x(k+1) \\
&\quad = \hat{x}(k/k) + f(\hat{x}(k/k))\,\Delta t - x(k) - f(x(k))\,\Delta t - G(t)w(k)\,\Delta t \\
&\quad \approx \tilde{x}(k/k) + F(\hat{x}(k/k))\tilde{x}(k/k)\,\Delta t - G(t)w(k)\,\Delta t;
\end{aligned} \tag{3.71}$$

then the covariance propagation equation is obtained:

$$
\begin{aligned}
\tilde{P}(k+1/k) &= E\{\tilde{x}(k+1/k)\tilde{x}(k+1/k)^T\} \\
&= E\{\tilde{x}(k/k)\tilde{x}(k/k)^T\} + E\{\tilde{x}(k/k)\tilde{x}(k/k)^T\}F(\hat{x}(k/k)^T\,\Delta t \\
&\quad + F(\hat{x}(k/k))E\{\tilde{x}(k/k)\tilde{x}(k/k)^T\}\,\Delta t \\
&\quad + E\{F(\hat{x}(k/k))F(\hat{x}(k/k))^T\}\,\Delta t^2 \\
&\quad + G(t)E\{w(k)w(k)^T\}G(t)^T\,\Delta t^2 \\
&= \tilde{P}(k/k) + \Big[F(\hat{x}(k/k))\tilde{P}(k/k) + \tilde{P}(k/k)F(\hat{x}(k/k))^T \\
&\qquad + G(t)Q(t)G(t)^T\Big]\,\Delta t + o(\Delta t). \qquad (3.72)
\end{aligned}
$$

Now dividing by Δt and taking the limit as $\Delta t \to 0$, $k\,\Delta t \to t$:

$$
d\tilde{P}/dt = F(\hat{x}(t/t))\tilde{P}(t) + \tilde{P}(t)F(\hat{x}(t/t))^T + G(t)Q(t)G(t)^T. \qquad (3.73)
$$

The covariance matrix $\tilde{P}(k+1/k+1)$ is given by

$$
\begin{aligned}
&\tilde{P}(k+1/k+1) \\
&\quad = E\{\tilde{x}(k+1/k+1)\tilde{x}(k+1/k+1)^T\} \\
&\quad = E\Big\{\Big[\tilde{x}(k/k) + F(\hat{x}(k/k))\tilde{x}(k/k)\,\Delta t \\
&\qquad + K(k+1)\big[z(k+1) - \hat{h}(x(k+1))\big] - G(t)w(t)\,\Delta t\Big] \\
&\qquad \times\Big[\tilde{x}(k/k) + F(\hat{x}(k/k))\tilde{x}(k/k)\,\Delta t + K(k+1) \\
&\qquad \times\big[z(k+1) - \hat{h}(x(k+1))\big] - G(t)w(t)\,\Delta t\Big]^T\Big\} \\
&\quad = \tilde{P}(k/k) + F(\hat{x}(k/k))\tilde{P}(k/k)\,\Delta t + \tilde{P}(k/k)F(\hat{x}(k/k))^T\,\Delta t \\
&\qquad + G(k)Q(k)\,\Delta t\, G(K)^T \\
&\qquad + E\Big\{(\tilde{x}(k/k) + \Delta f(\hat{x}(k/k))\,\Delta t - G(k)w(k)\,\Delta t) \\
&\qquad \times\big[z(k+1) - \hat{h}(x(k+1))\big]^T K(k+1)^T\,\Delta t\Big\} \\
&\qquad + E\Big\{K(k+1)\,\Delta t\big[z(k+1) - \hat{h}(x(k+1))\big](\tilde{x}(k/k) \\
&\qquad + \Delta f(\hat{x}(k/k))\,\Delta t - G(k)w(k)\,\Delta t]^T\Big\} \\
&\qquad + E\Big\{K(k+1)\big[z(k+1) - \hat{h}(x(k+1))\big] \\
&\qquad \times\big[z(k+1) - \hat{h}(x(k+1))\big]^T K(k+1)^T\Big\} \\
&\quad = \tilde{P}(k/k) + \Big[F(\hat{x}(k/k))\tilde{P}(k/k) + \tilde{P}(k/k)F(\hat{x}(k/k))^T \\
&\qquad + G(k)Q(k)\,\Delta t\, G(k)^T + E\{\tilde{x}(k/k)(z(k+1) \\
&\qquad - \hat{h}(x(k+1))^T K(t)^T\} + E\{K(t)(z(k+1) \\
&\qquad - \hat{h}(x(k+1))\tilde{x}(k/k)^T\}\Big]\,\Delta t + o(\Delta t),
\end{aligned}
$$

considering that

$$\begin{aligned}
\hat{h}(x(k+1)) &\approx h(\hat{x}(k+1/k)),\\
z(k+1) &= h(x(k+1)) + v(k+1)\\
&= h(\hat{x}(k+1/k)) - H(\hat{x}(k+1/k))\tilde{x}(k+1/k)\\
&\quad + \tfrac{1}{2}\tilde{x}(k+1/k)^T\,\partial^2 h(\hat{x}(k+1/k))/\partial x^2\\
&\quad \times \tilde{x}(k+1/k) + v(k+1),
\end{aligned}$$

$$\begin{aligned}
&z(k+1) - h(\hat{x}(k+1)/k)\\
&\quad = -H(\hat{x}(k+1/k))\tilde{x}(k/k)\\
&\qquad + \tfrac{1}{2}\tilde{x}(k/k)^T\,\partial^2 h(\hat{x}(k+1/k))/\partial x^2\tilde{x}(k/k) + v(k+1).
\end{aligned} \tag{3.74}$$

Substituting Eq. (3.74) into $\tilde{P}(k+1/k+1)$, and for $\hat{x}(k+1/k) = \hat{x}(k/k) + \Delta f\,\Delta t$,

$$\begin{aligned}
&\tilde{P}(k+1/k+1)\\
&\quad = \tilde{P}(k/k) + \Big[F(\hat{x}(k/k))\tilde{P}(k/k) + \tilde{P}(k/k)F(\hat{x}(k/k))^T\\
&\qquad + G(k)Q(k)\,\Delta t\,G(k)^T + E\Big\{\tilde{x}(k/k)\Big(-H(\hat{x}(k+1/k))\tilde{x}(k/k)\\
&\qquad + \tfrac{1}{2}\tilde{x}(k/k)^T\,\partial^2 h(\hat{x}(k+1/k))/\partial x^2\,\tilde{x}(k/k) + v(k+1)\Big)^T K(t)^T\Big\}\\
&\qquad + E\Big\{K(t)\Big(-H(\hat{x}(k+1/k))\tilde{x}(k/k)\\
&\qquad + \tfrac{1}{2}\tilde{x}(k/k)^T\,\partial^2 h(\hat{x}(k+1/k))/\partial x^2\,\tilde{x}(k/k) + v(k+1)\Big)\tilde{x}(k/k)^T\Big\}\\
&\qquad + E\{\tilde{x}(k/k)(-H(\hat{x}(k+1/k))\tilde{x}(k/k)\\
&\qquad + \tfrac{1}{2}\tilde{x}(k/k)\,\partial^2 h(\hat{x}(k+1/k))/\partial x^2\,\tilde{x}(k/k)\\
&\qquad + v(k+1))^T K(t)^T\Big]\,\Delta t + o(\Delta t)\\
&\quad = \tilde{P}(k/k) + \Big[F(\hat{x}(k/k))\tilde{P}(k/k) + \tilde{P}(k/k)F(\hat{x}(k/k))^T\\
&\qquad + G(k)Q(k)\,\Delta t\,G(k)^T - \tilde{P}(k/k)H(\hat{x}(k+1/k))^T K(t)^T\\
&\qquad - K(t)\hat{x}(k+1/k)\tilde{P}(k/k)\\
&\qquad + \tfrac{1}{2}E\Big\{\tilde{x}(k/k)\tilde{x}(k/k)^T\,\partial^2 h(\hat{x}(k+1/k))/\partial x^2\,\tilde{x}(k/k)\Big\}K(t)^T\Big\}\\
&\qquad + \tfrac{1}{2}E\Big\{K(t)\tilde{x}(k/k)^T\,\partial^2 h(\hat{x}(k+1/k))/\partial x^2\,\tilde{x}(k/k)\tilde{x}(k/k)^T\Big\}\\
&\qquad + K(t)R(k)\,\Delta t\,K(t)^T\Big]\,\Delta t + o(\Delta t).
\end{aligned} \tag{3.75}$$

Dividing by Δt, taking the limit as $\Delta t \to 0$, and $Q(k)\,\Delta t \to Q(t)$, $R(k)\,\Delta t \to R(t)$,

$$\begin{aligned}
d\tilde{P}(t/t)/dt \\
&= F(\hat{x}(t/t))\tilde{P}(t/t) + \tilde{P}(t/t)F(\hat{x}(t/t))^T + G(t)Q(t)G(t)^T \\
&\quad - \tilde{P}(t/t)H(\hat{x}(t/t))K(t)^T - K(t)H(\hat{x}(t/t))^T\tilde{P}(t/t) \\
&\quad + K(t)R(t)K(t)^T \\
&\quad + \tfrac{1}{2}E\{\tilde{x}(t/t)\tilde{x}(t/t)^T\,\partial^2 h(\hat{x}(t/t))/\partial x^2\,\tilde{x}(t/t)\}K(t)^T\} \\
&\quad + \tfrac{1}{2}E\{K(t)\tilde{x}(t/t)^T\,\partial^2 h(\hat{x}(t/t))/\partial x^2\,\tilde{x}(t/t)\tilde{x}(t/t)^T\}. \qquad (3.76)
\end{aligned}$$

Consider the gain that produces the minimum variance as before;

$$K(t) = \tilde{P}(t/t)H(x(t/t))^T R(t)^{-1}.$$

After some manipulations, as in McGarty (1974), Eq. (3.76) becomes

$$\begin{aligned}
d\tilde{P}(t/t)/dt \\
&= F(\hat{x}(t/t))\tilde{P}(t/t) + \tilde{P}(t/t)F(\hat{x}(t/t))^T + G(t)Q(t)G(t)^T \\
&\quad - \tilde{P}(t/t)H(\hat{x}(t/t))^T R(t)^{-1}H(\hat{x}(t/t))\tilde{P}(t/t) \\
&\quad + \tilde{P}(t/t)\sum_i \{\partial^2 h_i(\hat{x}(t/t))/\partial x^2 \\
&\qquad \times \gamma_i^T R(t)^{-1}[z(t) - h(\hat{x}(t/t))]\}\tilde{P}(t/t), \qquad (3.77)
\end{aligned}$$

where γ_i is the ith column of an $n \times n$ identity matrix I.

It is obvious from Eq. (3.77) that the continuous-time extended Kalman filter is not a direct extension of the linear case (see Table 3.4). The covariance matrix equation, in addition to the second derivatives of the measurement equation, depends on the real-time measurements. This makes the off-line calculation of the covariance matrix impossible. In summary,

$$\begin{aligned}
d\hat{x}(t/t)/dt &= f(\hat{x}(t/t)) + K(t)[z(t) - h(\hat{x}(t/t))], \qquad \hat{x}(0/0) = x_0, \\
K(t) &= \tilde{P}(t/t)H(\hat{x}(t/t))^T R(t)^{-1}, \qquad \tilde{P}(0/0) = P_0, \\
d\tilde{P}(t/t)/dt &= F(\hat{x}(t/t))\tilde{P}(t/t) + \tilde{P}(t/t)F(\hat{x}(t/t))^T + G(t)Q(t)G(t)^T \\
&\quad - \tilde{P}(t/t)H(\hat{x}(t/t))R(t)^{-1}H(\hat{x}(t/t))^T\tilde{P}(t/t) \\
&\quad + \tilde{P}(t/t)\sum_{i=1}^{n} \{\partial^2 h_i(\hat{x}(t/t))/\partial x^2 \\
&\qquad \times \gamma_i^T R(t)^{-1}[z(t) - h(\hat{x}(t/t))]\}\tilde{P}(t/t). \qquad (3.78)
\end{aligned}$$

TABLE 3.4 Summary of Continuous Extended Kalman Filter

System model	$dx/dt = f(t, x(t)) + G(t)w(t), \quad w(t) \sim N(0, Q(t))$
Measurement model	$z(t) = H(t)x(t) + v(t), \quad v(t) \sim N(0, R(t))$
Initial conditions	$E\{x(0)\} = \hat{x}_0, \quad E\{[x(0) - \hat{x}_0][x(0) - \hat{x}_0]^T\} = P_0$
Other assumptions	$R(t)^{-1} > 0$ exists, $\quad E\{w(t)v(\tau)^T\} = 0$
State estimate	$d\hat{x}/dt = f(t, \hat{x}(t)) + K(t)[z(t) - h(t, \hat{x}(t))]$, $\hat{x}(0) = \hat{x}_0$
Error covariance	$d\tilde{P}/dt = F(t, \hat{x}(t))\tilde{P}(t) + \tilde{P}(t)F(t, \hat{x}(t))^T + G(t)Q(t)G(t)^T - K(t)R(t)K(t) + \text{HOT}$, $\tilde{P}(0) = \tilde{P}_0$.
KB gain matrix	$K(t) = \tilde{P}(t)H(t, \hat{x}(t))^T R(t)^{-1}$
Definitions	$F(t, \hat{x}(t)) = \partial f(t, x(t))/\partial x \vert x(t) = \hat{x}(t)$ $H(t, \hat{x}(t)) = \partial h(t, x(t))/\partial x \vert x(t) = \hat{x}(t)$ HOT = higher-order terms

However, in the literature the continuous-time Kalman filter appears without the last term of the covariance matrix equation:

$$
\begin{aligned}
d\hat{x}(t/t)/dt &= f(\hat{x}(t/t)) + K(t)[z(t) - h(\hat{x}(t/t))], \qquad \hat{x}(0/0) = x_0, \\
K(t) &= \tilde{P}(t/t)H(\hat{x}(t/t))^T R(t)^T, \qquad \tilde{P}(0/0) = P_0, \\
d\tilde{P}(t/t)/dt &= F(\hat{x}(t/t))\tilde{P}(t/t) + \tilde{P}(t/t)F(\hat{x}(t/t))^T + G(t)Q(t)G(t)^T \\
&\quad - \tilde{P}(t/t)H(\hat{x}(t/t))R(t)^{-1}H(\hat{x}(t/t))^T\tilde{P}(t/t). \qquad (3.79)
\end{aligned}
$$

This approximation of the covariance matrix may contribute to instability of the solution and therefore divergence of the estimator.

As a solution, the continuous-time estimator is combined with a set of discrete-time observations, to produce the continuous-discrete extended Kalman filter, presented in the following. This formulation avoids the inclusion of the second-order terms, which are due to the Wiener–Lévy process, and thus makes the covariance matrix independent of the current measurements of the system. This filter is applicable with minimum danger of divergence, and it is very appealing for engineering applications.

3.12 CONTINUOUS-DISCRETE-TIME EXTENDED KALMAN FILTER

The problem of second derivative terms appearing in the covariance equation of the continuous-time extended Kalman filter are due to the measurement equation, and create considerable computational difficulties. To remedy that, it may be assumed that the measurements are performed at discrete-time instances, $z(k) = z(k\,\Delta t)$. Then although the estimate and covariance propagation equations remain in continuous time, as they are deduced from

Eqs. (3.66), (3.69), and (3.73),

$$d\hat{x}(t/t) = f(\hat{x}(t/t)), \qquad \hat{x}(0/0) = x_0,$$

$$d\tilde{P}(t/t)/dt = F(\hat{x}(t/t))\tilde{P}(t/t) + \tilde{P}(t/t)F(\hat{x}(t/t))^T + G(t)Q(t)G(t)^T,$$

$$\tilde{P}(0/0) = P_0, \tag{3.80}$$

the state estimate and error covariance update equations, deduced from Eqs. (3.66) and (3.59), at time $t = k\,\Delta t$, become

$$\begin{aligned}
\hat{x}(k+1/k+1) &= \hat{x}(k+1/k) + K(k+1)[z(k+1) - h(\hat{x}(k+1/k))], \\
\tilde{P}(k+1/k+1)) & \\
&= E\{\tilde{x}(k+1/k+1)\tilde{x}(k+1/k+1)^T\} \\
&= E\{[\tilde{x}(k+1/k) + K(k+1)[z(k+1) - h(\hat{x}(k+1/k))]] \\
&\qquad \times [\tilde{x}(k+1/k) + K(k+1)[z(k+1) - h(\hat{x}(k+1/k))]]^T\} \\
&= E\{[\tilde{x}(k+1/k) - K(k+1)H(\hat{x}(k+1/k))\tilde{x}(k+1/k)] \\
&\qquad \times [\tilde{x}(k+1/k) - K(k+1)H(\hat{x}(k+1/k))\tilde{x}(k+1/k)]^T\} \\
&= \tilde{P}(k+1/k) - K(k+1)H(\hat{x}(k+1/k))\tilde{P}(k+1/k), \\
K(k+1) &= \tilde{P}(k+1/k)H(\hat{x}(k+1/k))^T \\
&\qquad \times \left[H(\hat{x}(k+1/k)\tilde{P}(k+1/k)H(\hat{x}(k+1))^T + R(k)\right]^{-1},
\end{aligned} \tag{3.81}$$

where,

$$H(\hat{x}(k+1/k)) = \partial h(\hat{x}(k+1/k))/\partial x.$$

Summarizing the above equations, the continuous-discrete Kalman filter is given by (see Table 3.5)

$$\begin{aligned}
d\hat{x}(t/t) &= f(\hat{x}(t/t)), \qquad \hat{x}(0/0) = x_0, \\
d\tilde{P}(t/t)/dt &= F(\hat{x}(t/t))\tilde{P}(t/t) + \tilde{P}(t/t)F(\hat{x}(t/t))^T \\
&\qquad + G(t)Q(t)G(t)^T, \qquad \tilde{P}(0/0) = P_0, \\
\hat{x}(k+1/k+1) &= \hat{x}(k+1/k) + K(k+1)[z(k+1) - h(\hat{x}(k+1/k))], \\
\tilde{P}(k+1/k+1)) &= \tilde{P}(k+1/k) - K(k+1)H(\hat{x}(k+1/k))\tilde{P}(k+1/k), \\
K(k+1) &= \tilde{P}(k+1/k)H(\hat{x}(k+1/k))^T \\
&\qquad \times \left[H(\hat{x}(k+1/k)\tilde{P}(k+1/k)H(\hat{x}(k+1/k))^T + R(k)\right]^{-1}.
\end{aligned} \tag{3.82}$$

TABLE 3.5 Summary of Continuous-Discrete Extended Kalman Filter

System model	$dx/dt = f(t, x(t)) + G(t)w(t), \qquad w(t) \sim N(0, Q(t))$
Measurement model	$z(k) = h(k, x(k)) + v(k), \qquad v(k) \sim N(0, Q(k))$
Initial conditions	$E\{x(0)\} = \hat{x}_0, \qquad E\{[x(0) - \hat{x}_0][x(0) - \hat{x}_0]^T\} = P_0$
Other assumptions	$E\{w(k)v(j)\} = 0, \qquad \forall\, k, j.$
State estimate	$d\hat{x}/dt = f(t, \hat{x}(t)), \qquad \hat{x}(0) = \hat{x}_0$
Error covariance	$d\tilde{P}/dt = F(t, \hat{x}(t))\tilde{P}(t) + \tilde{P}(t)F(t, \hat{x}(t))^T$ $+ G(t)Q(t)G(t)^T, \qquad \tilde{P}(0) = \tilde{P}_0.$
Estimate update	$\hat{x}(k/k) = \hat{x}(k/k-1) + K(k)[z(k) - h(k, \hat{x}(k/k-1))]$
Covariance update	$\tilde{P}(k/k) = [I - K(k)H(k, \hat{x}(k/k-1))]\tilde{P}(k/k-1)$
KF gain matrix	$K(k) = \tilde{P}(k/k-1)H(k, \hat{x}(k/k-1))^T$ $\times[H(k, \hat{x}(k/k-1))\tilde{P}(k/k-1)H(k, \hat{x}(k/k-1))^T + R(k)]^{-1}$
Definitions	$F(t, \hat{x}(t)) = \partial f(t, x(t))/\partial x\, x(t) = \hat{x}(t)$ $H(k, \hat{x}(k/k-1)) = \partial h(k, x(k))/\partial x \vert x(k) = \hat{x}(k/k-1)$

The above equation gives satisfactory results in a large variety of estimation problems.

3.13 CONTINUOUS-TIME LINEARIZED KALMAN FILTER

If system (3.62) is linearized about a nominal trajectory $x'(t)$, as in the discrete-time case, the second derivative terms do not appear in the error covariance matrix anymore, because the estimator is linear. The continuous-time linearized Kalman filter equations can be derived in a manner analogous to Eqs. (3.61):

$$d\hat{x}(t/t)/dt = f(x'(t)) + F(\hat{x}(t))(\hat{x}(t/t)) - x'(t))$$
$$+ K(t)[z(t) - h(x'(t)) - H(x'(t))(\hat{x}(t/t) - x'(t))],$$
$$(0/0) = x_0,$$

$$K(t) = \tilde{P}(t/t)H(x'(t))R(t)^{-1},$$

$$d\tilde{P}(t/t)/dt = F(x'(t))\tilde{P}(t/t) + \tilde{P}(t/t)F(x'(t))^T + G(t)Q(t)G(t)^T$$
$$-\tilde{P}(t/t)H(x'(t))^T R(t)^{-1} H(x'(t))\tilde{P}(t/t),$$
$$\tilde{P}(0/0) = P_0. \quad (3.83)$$

3.14 CONTINUOUS-DISCRETE-TIME LINEARIZED KALMAN FILTER

A continuous-discrete version of the linearized Kalman filter popular in the literature is given by (see Table 3.6)

$$
\begin{aligned}
d\hat{x}(t/t)/dt &= f(x'(t)) + F(x'(t))(\hat{x}(t/t)) - x'(t)), \qquad \hat{x}(0/0) = x_0, \\
d\tilde{P}(t/t)/dt &= F(x'(t))\tilde{P}(t/t) + \tilde{P}(t/t)F(x'(t))^T \\
&\quad + G(t)Q(t)G(t)^T \tilde{P}(0/0) = P_0, \\
\hat{x}(k+1/k+1) &= \hat{x}(k+1/k) + K(k+1)[z(k+1) - h(x'(k+1)) \\
&\quad - H(x'(k))(\hat{x}(k+1/k) - x'(k+1))], \\
\tilde{P}(k+1/k+1)) &= \tilde{P}(k+1/k) - K(k+1)H(x'(k+1))\tilde{P}(k+1/k), \\
K(k+1) &= \tilde{P}(k+1/k)H(x'(k+1))^T \\
&\quad \times \left[H(x'(k+1)\tilde{P}(k+1/k)H(x'(k+1))^T + R(k)\right]^{-1}.
\end{aligned}
\tag{3.84}
$$

This filter is popular due to its similarity with the continuous-discrete-extended Kalman filter, even though the original continuous-time linearized Kalman filter does not present the same problems. It is quite often used in appropriate engineering applications.

TABLE 3.6 Summary of Continuous-Discrete Linearized Kalman Filter

System model	$dx/dt = f(t, x(t)) + G(t)w(t), \quad w(t) \sim N(0, Q(t))$
Measurement model	$z(k) = h(k, x(k)) + v(k), \quad v(k) \sim N(0, Q(k))$
Initial conditions	$E\{x(0)\} = \hat{x}_0, \quad E\{[x(0) - \hat{x}_0][x(0) - \hat{x}_0]^T\} = P_0$
Other assumptions	$E\{w(k)v(j)\} = 0, \quad \forall\, k, j;\ x'(t) =$ nominal trajectory
State estimate	$d\hat{x}/dt = f(t, x'(t)) + F(t, x'(t))[\hat{x}(t) - x'(t)], \quad \hat{x}(0) = \hat{x}_0$
Error covariance	$d\tilde{P}/dt = F(t, x'(t))\tilde{P}(t) + \tilde{P}(t)F(t, x'(t))^T$ $+ G(t)Q(t)G(t)^T, \quad \tilde{P}(0) = \tilde{P}_0.$
Estimate update	$\hat{x}(k/k) = \hat{x}(k/k-1) + K(k)[z(k) - h(k, x'(k/k-1))]$
Covariance update	$\tilde{P}(k/k) = [I - K(k)H(k, x'(k/k-1))]\tilde{P}(k/k-1)$
KF gain matrix	$K(k) = \tilde{P}(k/k-1)H(k, x'(k/k-1))^T$ $\times [H(k, x'(k/k-1))\tilde{P}(k/k-1)H(k, x'(k/k-1))^T + R(k)]^{-1}$
Definitions	$F(t, x'(t)) = \partial f(t, x(t))/\partial x \vert x(t) = x'(t)$ $H(k, x'(k/k-1)) = \partial h(k, x(k))/\partial x \vert x(k) = x'(k/k-1)$

PROBLEMS

3.1 Using the recursive estimator developed in the previous chapter, and the best estimate of the time propagation of the state, derive the best state estimator for the scalar discrete-time system:

$$x(k+1) = x(k) + w(k),$$
$$z(k) = x(k) + v(k),$$

where $w(k) \sim N(0, \sigma^2)$, $v(k) \sim N(0, \alpha^2)$, $x(0) \sim N(0, \gamma^2)$ the system random variables. Give explicit values of the solution.

3.2 Given the discrete-time stochastic system

$$x_1(k+1) = x_1(k) + x_2(k),$$
$$x_2(k+1) = x_2(k) + x_3(k),$$
$$x_3(k+1) = 0.5x_1(k) - 1.5x_2(k) + x_3(k) + w(t),$$
$$z(t) = x_1(k) + v(t),$$

where $w(t) \sim N[0, \sigma^2]$, $v(t) \sim N[0, \sigma^2]$, are the process and measurement noises, respectively, and $x(0) = x_0 \sim N[0, \Sigma^2]$ is the initial condition vector. Derive the equations of the discrete Kalman filter estimator and draw a block diagram to show its implementation.

3.3 Show that for the discrete-time problem with an additional deterministic control input $u(t)$ and system equations

$$x(k+1) = Fx(k) + Bu(k) + Gw(k), \qquad x(0) \sim N(x_0, P_0)$$
$$z(k) = h^T x(k) + v(k), \qquad w(k) \sim N(0 \cdot q^2), v(k) \sim N(0, r^2),$$

and the equations that derive the discrete-time Kalman filter:

$$\hat{x}(k/k) = \hat{x}(k-1/k-1)$$
$$-\{\tilde{P}(k/k-1)h[z(k) - h^T\hat{x}(k/k-1)\}/\{h^T\tilde{P}(k/k-1)h + r^2\},$$

$$\tilde{P}(k/k) = \tilde{P}(k/k-1)$$
$$-\{\tilde{P}(k/k-1)hh^T\tilde{P}(k/k-1)\}/\{h^T\tilde{P}(k/k-1)h + r^2\},$$

the control term of the Kalman filter reads

$$\tilde{P}(k/k)\tilde{P}(k/k-1)^{-1}Bu(k).$$

3.4 Given the continuous-time stochastic system:

$$dx/dt = -0.5x(t) = w(t),$$
$$z(t) = x(t).$$

Find the continuous-time Kalman–Bucy filter, when $x(0) \sim N(0, p_0,$ $w(t) \sim N(0, \sigma^2)$. Give explicit values for the solution of the Riccati equation.

3.5 Given the continuous-time scalar stochastic system:

$$dx/dt = -x(t),$$
$$z(t) = x(t) + v(t),$$

obtain the continuous-time Kalman–Bucy filter, when $x(0) \sim N(0, p_0)$, $v(t) \sim N(0, r^2)$, and the solution of the variance equation.

3.6 For the continuous-time system with fixed initial conditions x_0,

$$dx/dt = Fx(t) + Gw(t),$$
$$z(t) = Hx(t) + v(t),$$

with

$$w(t) \sim N(0, Q\,\delta(t-\tau)), v(k) \sim N(0, R\,\delta(t-\tau)),$$

consider a linear filter described by

$$d\hat{x}/dt = K_1\hat{x}(t/t) + K_2 z,$$

where K_1 and K_2 are to be chosen to optimize the estimate $x(t/t)$. By requiring that the estimate be unbiased, show that $K_1 = F - KH$, thus obtaining the estimator

$$d\hat{x}\,dt = F\hat{x}(t/t) + K_2[z - H\hat{x}(t/t)].$$

Then show that the covariance equation of the estimation error is

$$d\tilde{P}/dt = (F - KH)\tilde{P} + \tilde{P}(F - KH)^T + GQG^T + KRK^T$$

Finally, choose K_2 to yield a maximum rate of decrease of error by minimizing the cost function

$$J = \text{Trace}\left[d\tilde{P}/dt\right],$$

which corresponds to the Kalman gain $K_2 = \tilde{P}H^TR^{-1}$.
[Note: Use the properties

$$d/dA\left[\text{Trace}(ABA^T)\right] = 2AB, d\left[\text{Trace}(AB^T)\right]/dA = B.]$$

REFERENCES

Aoki, M., *Introduction to Stochastic Systems*, Academic Press, New York, 1967.

Deutz, R., *Estimation Theory*, Prentice Hall, Englewood Cliffs, NJ, 1965.

Gelb, A., *Applied Optimal Estimation*, MIT Press, Cambridge, MA, 1974.

Itô, K., *On Stochastic Differential Equations*, *Mem. Am. Math. Soc.*, Vol. 4, 1951.

Kalman, R. E., "A New Approach to Linear Filtering and Prediction" *ASME Trans. J. Basic Eng.*, Vol. 82, pp. 34–35 (1960).

Kalman, R. E., "When is a Linear Control Optimal?" *ASME Trans. J. Basic Eng.*, Vol. 86, pp. 51–60 (1964).

Kalman, R. E., and Bucy R., "New Results in Linear Filtering and Prediction Theory," *ASME Trans. J. Basic Eng.* pp. 83–95 (1961).

Lee, R. C. K., *Optimal Identification, Estimation, and Control*, MIT Press, Cambridge, MA, 1964.

Lewis, F. L., *Optimal Estimation: Introduction to Stochastic Control*, Wiley-Interscience, New York, 1986.

McGarty, T. P., *Stochastic Systems and State Estimation*, Wiley-Interscience, New York, 1974.

Meditch, J., *Stochastic Linear Estimation and Control*, McGraw-Hill, New York, 1969.

Sage, A. P., and Melsa, J. L., *System Identification*, Academic Press, New York, 1971.

Saridis, G. N., *Self-Organizing Controls of Stochastic Systems*, Marcel Dekker, New York, 1977.

Sorenson, H. W., *Parameter Estimation*, Marcel Dekker, New York, 1980.

4

STOCHASTIC OPTIMAL CONTROL

4.1 STOCHASTIC OPTIMAL-CONTROL PROBLEM

The stochastic control problem arises when uncertainties are present about the initial position, the process, and the measurements of a system to be controlled. These uncertainties may be modeled by stochastic processes, with probability distributions as their measure. Then the problem is mapped onto the probability space and solved deterministically. The result is usually a probabilistic measure of the response of the system expressing the certainty (or uncertainty) of the values of the response. Such a result is, in most cases, impossible to obtain due to the limitations of the mathematical operations required; it is also impractical for most technical as well as other applications.

As with the estimation problem, the stochastic control problem is simplified by considering the statistical moments of its behavior. The expected value gives the most likely value, in some sense, and the covariance represents our confidence in that value. To avoid unnecessary complex calculations, higher moments are not usually considered. Such results have been strongly criticized by practicing engineers because of the nondeterministic nature of the results. However, in an environment dominated by uncertainties, either due to our own ignorance (subjective), or due to the unpredictability of Mother Nature, the stochastic model produces the most satisfactory global results and relates the generated solutions to generic physical quantities, for example, entropy.

Optimal control is one possible manifestation of the solution of the control problem. If the control problem is defined as finding the method with which a system would behave according to specifications, and/or tracking some desired behavior, optimal control attains this goal by finding a certain control function (law) that optimizes a cost functional containing all the desired specifications. When the system is modeled stochastically, the control function should be a deterministic function of the system's stochastic states and parameters and their covariances. Stochastic control functions (strategies) are also possible for some problems, and they are considered in *dual control*. However, such solutions have found only theoretical acclaim, because their technical application is usually impractical. The stochastic optimal-control formulation will be presented, first for the discrete-time and then for the continuous-time cases, following the approach that was pursued for the estimation problem, in order to avoid unnecessary complex mathematical derivations.

First, the analytic formulation of nonlinear stochastic optimal control will be presented, where the assumption is made that both the process and the measurement uncertainty (noise) are independent and identically distributed (i.i.d.) stochastic processes that are uncorrelated with the random initial position of the states of the system, and enter the system linearly.

4.1.1 Continuous-Time System

The system

$$dx/dt = f(x, u, t) + G(t)w(t), \qquad x(0), \tag{4.1}$$

$$z(t) = h(x, u, t) + v(t),$$

is given, where $x(t)$ is the state of the system at time t, $z(t)$ is the current m-dimensional measurement vector, $u(t)$ is the r-dimensional control vector, $f(x, u, t), h(x, u, t)$ are twice-differentiable nonlinear functions of their arguments, and $G(t)$ is an $n \times r$ matrix of known coefficients, respectively. The stochastic variable $x(0)$ is the n-dimensional Gaussian initial state, $w(t)$ is the r-dimensional process noise, $v(t)$ is the m-dimensional measurement noise, defined for any $t \in [0, \infty]$, with properties

$$E\{x(0)\} = x_0, \qquad E\{(x(0) - x_0)(x(0) - x_0)^T\} = P_0,$$

$$E\{w(t)\} = 0, \qquad E\{w(t)w(\tau)^T\} = Q(t)\,\delta(t - \tau),$$

$$E\{v(t)\} = 0, \qquad E\{v(t)v(\tau)^T\} = R(t)\,\delta(t - \tau),$$

$$E\{x(0)w(t)^T\} = E\{x(0)v(t)^T\} = E\{w(t)v(\tau)^T\} = 0. \tag{4.2}$$

A cost functional containing all the predefined desired specifications about the system's performance is also defined:

$$I(u) = E\{J(u)\} = E\left\{\phi(x(T),T) + \int_0^T L(x,u,\tau)\,d\tau\right\}. \tag{4.3}$$

The problem is to find a u^*:

$$I(u^*) = I^* = \min_u I(u).$$

4.1.2 Discrete-Time System

$$\begin{aligned} x(k+1) &= f(x(k),u(k),k) + G(k)w(k), \qquad x(0), \\ z(k) &= h(x(k),u(k),k) + v(k), \end{aligned} \tag{4.4}$$

where $z(k)$ is a measurement m vector, $x(k)$ is a state n vector, $u(k)$ is a control r vector, and $f(\cdot,\cdot)$, $h(\cdot,\cdot)$ are nonlinear twice-differentiable functions, respectively. The stochastic variables, $x(0)$, the n-dimensional initial state, $w(k)$, an r-dimensional process noise, and $v(k)$, an m-dimensional measurement noise, have the following properties for any time $k \in [0, N]$, and δ_{ij} the Kronecker delta:

$$\begin{aligned} &E\{x(0)\} = x_0, \quad E\{(x(0)-x_0)(x(0)-x_0)^T\} = P_0, \\ &E\{w(k)\} = 0, \quad E\{w(k)w(j)^T\} = Q(k)\,\delta_{kj}, \quad E\{w(k)x(0)^T\} = 0, \\ &E\{v(k)\} = 0, \quad E\{v(k)v(j)^T\} = R(k)\,\delta_{kj}, \quad E\{v(k)x(0)^T\} = 0, \\ &E\{x(0)w(k)^T\} = E\{x(0)v(k)^T\} = E\{w(k)v(j)^T\} = 0. \end{aligned} \tag{4.5}$$

The performance cost functional is similarly defined:

$$I(u) = E\{J(u)\} = E\left\{\phi(x(N),N) + \sum_{i=0}^{N-1} L(x(i+1),u(i),i)\right\}. \tag{4.6}$$

The control problem is to find a control function u^*:

$$I(u^*) = I^* = \min_{\{u(k)\}} I(u(k)). \tag{4.7}$$

The discrete-time case will be pursued first, and the continuous-time case will be treated as the limit as $\Delta t \to 0$. Therefore, the following definitions and principles will be developed in discrete time. Their generalization to continuous time is straightforward.

4.1.3 Definitions: The Stochastic Optimal-Control Solutions

In contrast to the deterministic case, stochastic optimal-control accepts various solutions, not equivalent to each other. This has created considerable confusion in the literature. The following three optimal solutions *are ordered according to the decreasing value of their cost function*, and depend on the use of the information available.

Open-Loop Control

$$u^*(p(x_0), p(w), p(v), k): J(u^*) = \min_{\{u\}} E\left\{ \sum_{i=0}^{N-1} L(x(i+1), u(i), i) \right\}. \tag{4.8}$$

Here the optimization is performed off line, before starting the process, and therefore the controller depends only on the a priori information on the initial condition $p(x_0)$, process noise $p(w(k))$, and measurement noise $p(v(k))$. It has the highest cost of all the optimal solutions, as will be later demonstrated.

Passive Feedback (Open-Loop-Feedback-Optimal) Control

$$u^*(x(k), k, Z^k, U^{k-1}): J(u^*)$$
$$= \min_{\{u\}_k^{N-1}} E\left\{ \sum_{i=k}^{N-1} L(x(i+1), u(i), i)/Z^k, U^{k-1} \right\}. \tag{4.9}$$

The optimization in this case is performed at time k, where the controller is using the complete information on the past measurements Z^k, and control actions U^{k-1}, but minimizes open loop over the rest of the process. This is actually a pseudo-optimal solution, because the controller does not anticipate reduction of the future uncertainties through the control action. The only time that this solution is fully optimal is when no reduction of the future uncertainties is taking place, as in the case of linear-quadratic-gaussian (LQG) systems. Such systems were called *neutral* by Fel'dbaum (1965).

Active Feedback (Dual) Control

$$u^*(x(k), k, Z^k, U^k): V_{N-k}(x(k))$$
$$= \min_{u(k)} E\{L(x(k+1), u(k), k) + V_{N-k-1}(x(k+1))/Z^k U^{k-1}\}. \tag{4.10}$$

Active anticipation of reduction of the future uncertainties at time k, in addition to the use of all past measurements Z^k, and control actions U^{k-1}, makes this controller the true optimal controller with cost minimum compared to that previously defined. It assumes that the controlled solution is backtracked from the end of the process when the measurement and control information is applied. Fel'dbaum (1965) formulated this solution using Bellman's dynamic programming, to obtain the true optimal solution. His formulation of the dual control problem resulted in the conclusion that the solution of the nonlinear case is not always possible, even with the assistance of the modern superfast digital computers. His results are still not completely understood by many authors in the literature, who are confusing the optimal active feedback optimality with the open-loop-feedback-optimal solution.

The solution of the linear-quadratic-gaussian case that was first discovered by Tou and Joseph (1961), for discrete-time systems and generalized by Wonham, Bellman, and others, produced the so-called *separation theorem*, which yielded an optimal solution by cascading a Kalman filter with a deterministic optimal control. This was too good to be true as a solution, and was immediately generalized for all types of systems without much scrutiny (certainty equivalence principle). The result was a claim of optimality to many stochastic control problems that were simply open-loop-feedback-optimal (OLFO).

In the case that the systems defined above are Markovian processes,

$$p(y(k)/Z^k) = p(y(k)/z(k)); \tag{4.11}$$

Z^k, U^k, may be replaced by their sufficient statistics $z(k), u(k)$.

4.2 BELLMAN'S DYNAMIC PROGRAMMING EQUATION

Consider a process starting at $x(0)$ and driven to $x(N)$ by a control sequence $\{u(k)\}$. See Figure 4.1. Using the *principle of invariant imbedding*, the starting point may be moved to an arbitrary but fixed state $x(i)$. Define the cost of the $N - i$-stage process to be executed, starting at $x(i)$ and ending at $x(N)$, as $V_{N-i}(x(i))$. The cost of the process starting at $x(0)$ under the control

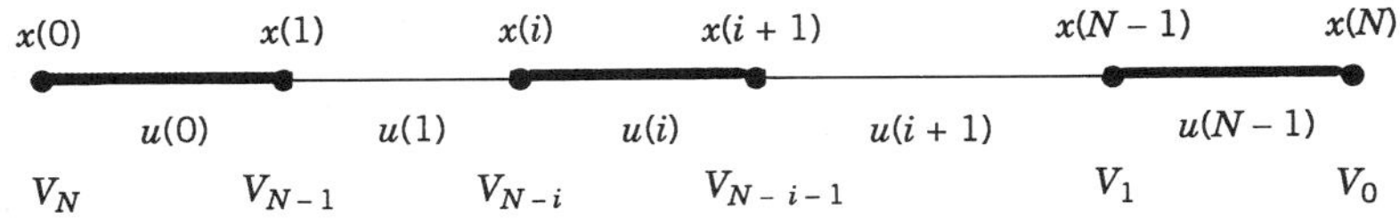

Fig. 4.1 Bellman's N-stage process for dynamic programming

sequence $\{u\} \in \Omega_u$ is given by its performance cost in Eqs. (4.6) and (4.7):

$$V_N(x(0)) = \min_{\{u\}} E\left\{\phi(x(N), N) + \sum_{i=0}^{N-1} L(x(i+1), u(i), i)\right\}. \quad (4.12)$$

4.2.1 Bellman's Principle of Optimality

An *optimal-control policy at state* $x(i)$ *leading to the end of the process must constitute an optimal policy regardless of the state* $x(i)$ *and the previous controls*.

A formal proof is given in Figure 4.1 by considering the alternate path leading to $x'(N)$, for which $V_0(x'(N)) > V_0(x(N))$.

4.2.2 The One-Stage Process

Suppose that the process under consideration has only one stage from $x(N-1)$ to $x(N)$. If the exit cost from stage $x(N)$ of system (4.6) is $V_0(x(N))$, the cost at $x(N-1)$ depending only on $u(N-1) \in \Omega_u$ is

$$V_0(x(N)) = \phi(x(N), N)$$

$$\begin{aligned} V_1(x(N-1)) &= \min_{\{u\}} E\{L(x(N), u(N-1), N-1) + V_0(x(N))\} \\ &= \min_{\{u\}} E\{L(f(x(N-1), u(N-1), N-1), u(N-1), N-1) \\ &\qquad + \phi(x(N-1), u(N-1), N-1)\}. \end{aligned} \quad (4.13)$$

The above expression yields two equations: one for the optimal control $u(N-1)^*$ and one for the optimal cost $V_1(x(N-1))^*$, after substituting $u(N-1)^*$ into Eq. (4.13).

4.2.3 The $N - i$-Stage Process

This is the general case when i is arbitrary but fixed. The process starting at $x(i)$ and ending at $x(N)$ has a cost

$$
\begin{aligned}
V_{N-i}(x(i)) &= \min_{u(i)\cdots u(N-1)} E\left\{\sum_{k=i}^{N-1} L(x(k+1), u(k), k) + \phi(x(N))\right\} \\
&= \min_{u(i)\cdots u(N-1)} E\left\{L(x(i+1), u(i), i) + \sum_{k=i+1}^{N-1} L(x(k+1),\right. \\
&\qquad\qquad \left. u(k), k) + \phi(x(N))\right\} \\
&= \min_{u(i)} E\left\{L(x(i+1), u(i), i) + \min_{u(i+1)\cdot u(N-1)} \sum_{k=i+1}^{N-1} L(x(k+1),\right. \\
&\qquad\qquad \left. u(k), k) + \phi(x(N))\right\} \\
&= \min_{u(i)} E\{L(x(i+1), u(i), i) + V_{N-i-1}(x(i+1))\}.
\end{aligned}
\tag{4.14}
$$

Expression (4.14) yields again two equations: one for the optimal control $u(i)^*$ and one for the optimal cost $V_{N-i}(x(i))^*$, after substituting $u(i)^*$ into Eq. (4.14).

The total optimal cost of the original process defined by Eq. (4.6) is then

$$
V_N(x(0)) = \min_{u(0)} E\{L(x(1), u(0), 0) + V_{N-1}(x(1))\}. \tag{4.15}
$$

Complete State Information. The above derivation was made with the assumption that all the states were available for measurement. Given the previous measurements of the states X^i, Eq. (4.14) can be written

$$
\begin{aligned}
&V_{N-i}(x(i)/X^i) \\
&\quad = \min_{u(i)} E\{L(x(i+1), u(i), i) + V_{N-i-1}(x(i+1)/X^{i+1})/X^i\}.
\end{aligned}
$$

Incomplete State Information. If the measurements $Z_0^{N-1} = \{z(0) \cdots z(N-1)\}$ were available as the optimal solution was obtained by backtracking the system from its end point, an active feedback control law $u^*(x(i), i)$ is obtained, conditioned on these measurements as guaranteed by the orthogonal projection theorem. In order to consider the measurements at time k, define

$$V_{N-i}(x(i)/Z^i) = \min_{u(i)} E\{L(x(i+1), u(i), i) + V_{N-i-1}(x(i+1))/Z^i\}. \tag{4.16}$$

Under this definition Eq. (4.14) becomes

$$E_{Zi}\{V_{N-i}(x(i)/Z^i)\}$$
$$= \min_{u(i)} E_{Zi}\{E\{L(x(i+1), u(i), i) + V_{N-i-1}(x(i+1))/Z^i\}\},$$

or by considering the minimization for all possible sequences Z^i,

$$V_{N-i}(x(i)/Z^i)$$
$$= \min_{u(i)} \{E\{L(x(i+1), u(i), i) + V_{N-i-1}(x(i+1)/Z^{i+1})/Z^i\}\}. \tag{4.17}$$

This is Bellman's equation of dynamic programming for stochastic optimal-control systems. It provides active feedback control because it starts the minimization procedure at the end of the process and moves backward, thus assuming control activity along the entire process. The resulting control is globally optimal, a fact that will be analytically established in Chapter 6. Equation (4.17) applies to all linear or nonlinear dynamic systems. However, its solution is rarely analytically or computationally feasible, due to the multiboundary value conditions arising from its formulation. This problem was recognized by Fel'dbaum in his work on Dual Control (1965).

4.3 CONTINUOUS-TIME HAMILTON–JACOBI–BELLMAN EQUATION

A continuous-time version of Bellman's equation (4.17) can be developed for the stochastic system of Eqs. (4.1)–(4.3). It resembles the deterministic Hamilton–Jacobi equation, but contains an additional term due to the imbedded Wiener process. A procedure that considers a series expansion before the limit, similar to the derivation of the continuous Kalman filter, makes the system discrete time. The equation will be derived for all the states that are available for measurement, from a set X^t, the sigma algebra of $x(t)$. It should be remembered that under those circumstances

$$E\{w(t)\} = 0, \qquad E\{w(t)w(\tau)^T\} = Q(t)\,\delta(t-\tau) = \lim_{\Delta t\to 0} \frac{Q(t)}{\Delta t}\delta_{t,t+\Delta t},$$
$$E\{v(t)\} = 0, \qquad E\{v(t)v(\tau)^T\} = R(t)\,\delta(t-\tau) = \lim_{\Delta t\to 0} \frac{R(t)}{\Delta t}\delta_{t,t+\Delta t}. \tag{4.18}$$

Complete State Information. Define the function $V(t/X^t)$ as follows

$$V(t/X^t) = \min E\{J(u)/X^t\} = E\left\{\phi(x(T),T) + \int_t^T L(x,u^*,\tau)\,d\tau/X^t\right\}; \tag{4.19}$$

then

$$dV/dt = -E\{L(x,u(x,t)^*,t)/X^t\} = -L(x,u^*,t). \tag{4.20}$$

Consider now Eq. (4.20) as a limiting process of a discrete-time system, with time increments $\Delta t \to 0$. Then

$$\begin{aligned}
dV/dt &= \lim_{\Delta t\to 0}\left[(V(t+\Delta t) - V(t))/\Delta t\right]\\
&= E\Big\{\lim_{\Delta t\to 0} 1\big/\Delta t\big[V(t) + \partial V/\partial t\,\Delta t + \partial V^T/\partial x\,dx/dt\,\Delta t + \tfrac{1}{2}\partial^2 V/\partial t^2\,\Delta t^2\\
&\qquad + \tfrac{1}{2}\,dx/dt^T\,\partial^2 V/\partial x^2\,dx/dt\,\Delta t^2 + o(\Delta t^2) - V(t)\big]/X^t\Big\}\\
&= \lim_{\Delta t\to 0} E\Big\{\partial V/\partial t + \partial V^T/\partial x f(x,u^*,t) + \partial V^T/\partial x\,G(t)w(t)\\
&\qquad + \tfrac{1}{2}\partial^2 V/\partial t^2\,\Delta t + \tfrac{1}{2} f(x,u^*,t)^T\,\partial^2 V/\partial x^2\,f(x,u^*,t)\,\Delta t\\
&\qquad + f(x,u^*,t)^T\,\partial^2 V/\partial x^2\,G(t)w(t)\,\Delta t\\
&\qquad + \tfrac{1}{2}\,\mathrm{Tr}\big[\partial^2 V/\partial x^2\,G(t)ww^T G(t)^T\big]\,\Delta t + o(\Delta t)/X^t\Big\}\\
&= \lim_{\Delta t\to 0}\Big[\partial V/\partial t + \partial V^T/\partial x f(x,u^*,t) + \tfrac{1}{2}\partial^2 V/\partial t^2\,\Delta t\\
&\qquad + \tfrac{1}{2} f(x,u^*,t)^T\,\partial^2 V/\partial x^2\,f(x,u^*,t)\,\Delta t\\
&\qquad + \tfrac{1}{2}\,\mathrm{Tr}\big[\partial^2 V/\partial x^2\,G(t)Q(t)/\Delta t\,G(t)^T\big]\,\Delta t + o(\Delta t)\Big]\\
&= \partial V/\partial t + \partial V^T/\partial x f(x,u^*,t) + \tfrac{1}{2}\,\mathrm{Tr}\big[\partial^2 V/\partial x^2\,G(t)Q(t)G(t)^T\big]\\
&= -L(x,u^*,t).
\end{aligned}$$

Finally, by rearranging the last expression, the *Hamilton–Jacobi–Bellman* equation results for stochastic control systems:

$$\partial V/\partial t + \min_u \Big[\partial V^T/\partial x f(x,u,t) + L(x,u,t)$$
$$+ \tfrac{1}{2}\operatorname{Tr}\big[\partial^2 V/\partial x^2\, G(t)Q(t)G(t)^T\big]\Big] = 0,$$
$$V(T/X^T) = \phi(x(T),T). \tag{4.21}$$

The equation is similar to the Hamilton–Jacobi equation for deterministic systems, except for the last term, which is the result of the stochastic differentiation of the Wiener–Lévy processes.

Incomplete State Information. In the case that only the measurements Z^t are available, one may replace $V(t/X^t)$ with $V(t/Z^t)$, and proceed with Eq. (4.21), as with the discrete-time case.

The above equations apply to general nonlinear continuous and twice-differentiable systems. They meet the requirements of the formulation of the stochastic optimal control with active transmission of information, derived by Fel'dbaum, because they start the minimization at the end of the process, thus assuming control activity along the whole process. Such a formulation yields a globally optimal solution, stated in words by *dual control*, but analytically proven only in Chapter 6 of the present work. Solution of the Hamilton–Jacobi–Bellman equation is not possible in most cases of nonlinear systems, as claimed by Fel'dbaum (1965). Modern computational techniques may eventually resolve some of these problems. However, engineers still prefer approximation methods of their execution like the one proposed in the next chapter, because approximation methods are simple to execute.

4.4 LINEAR-QUADRATIC-GAUSSIAN OPTIMAL-CONTROL PROBLEM: THE DISCRETE-TIME CASE

Optimal control will be derived for the linear system quadratic cost function and Gaussian random noise sequences (LQG). For such a case the system is neutral, and the solution can be obtained in a straightforward manner. The resulting active feedback control is the deterministic control solution applied on the optimal estimates of the states, obtained from a Kalman filter. This result is called the *separation theorem*, and the LQG case is its only well-known application.

4.4.1 Complete State Information

As with state estimation the discrete-time case will be considered first, and the continuous-time case will be obtained as its limit. Consider the linear time-varying system

$$\begin{aligned} x(k+1) &= F(k)x(k) + B(k)u(k) + G(k)w(k), \qquad x(0), \\ z(k) &= H(k)x(k) + v(k), \end{aligned} \tag{4.22}$$

where $z(k)$ is a measurement m vector, $x(k)$ is a state n vector, $u(k)$ is a control r vector, and $F(k)$, $B(k)$, $G(k)$, and $H(k)$, are $n \times n$, $n \times r$, $n \times r$, and $m \times n$ matrices, respectively. The stochastic variables, $x(0)$, the n-dimensional initial state, $w(k)$, an r-dimensional process noise, and $v(k)$, an m-dimensional measurement noise, have the following properties for any time $k \in [0, N]$:

$$\begin{aligned} &E\{x(0)\} = x_0, \qquad E\{(x(0) - x_0)(x(0) - x_0)^T\} = P_0, \\ &E\{w(k)\} = 0, \qquad E\{w(k)w(j)^T\} = Q(k)\,\delta_{kj}, \qquad E\{w(k)x(0)^T\} = 0, \\ &E\{v(k)\} = 0, \qquad E\{v(k)v(j)^T\} = R(k)\,\delta_{kj}, E\{v(k)x(0)^T\} = 0, \\ &E\{x(0)w(k)^T\} = E\{x(0)v(k)^T\} = E\{w(k)v(j)^T\} = 0. \end{aligned} \tag{4.23}$$

The quadratic performance cost functional is defined, for $\phi(x(N)) = 0$,

$$\begin{aligned} I(u) &= E\{J(u)\} \\ &= E\left\{ \sum_{i=0}^{N-1} \left[x(i+1)^T M(i+1) x(i+1) + u(i)^T N(i) u(i) \right] \right\}. \end{aligned} \tag{4.24}$$

The control problem is to find a control function u^*:

$$I(u^*) = I^* = \min_{\{u(k)\}} I(u(k)). \tag{4.25}$$

Using Bellman's equation (4.17), the following expression is obtained:

$$\begin{aligned} &V_{N-k}(x(k)/X^k) \\ &\quad = \min_{u(k)} \Big[E\big\{ x(k+1)^T M(k+1) x(k+1) + u(k)^T N(k) u(k) \\ &\qquad\qquad + V_{N-k-1}(x(k+1)/X^{k+1})/X^k \big\} \Big] \\ &\quad \triangleq E\{x(k)^T S_{N-k}(k) x(k)/X^k\} + q_{N-k}, \end{aligned} \tag{4.26}$$

where the assumption was made that the V function takes a quadratic form. Because all the other expressions are quadratic, Eq. (4.26) is reproducible for all k, and therefore its solution is possible:

$$E\{x(k)^T S_{N-k}(k)x(k)/X^k\} + q_{N-k}$$
$$= \min_{u(k)} \Big[E\{x(k+1)^T M(k+1)x(k+1) + u(k)^T N(k)u(k)$$
$$+x(k+1)^T S_{N-i-1}(k+1)x(k+1)/X^k\}\Big]. \quad (4.27)$$

Minimizing Eq. (4.27) by taking partials with respect to $u(k)$, and writing $S_{N-k}(k)$ as S_{N-k}, results in

$$u^*(k) = -\big[B(k)^T(M(k+1) + S_{N-k-1})B(k) + N(k)\big]^{-1}$$
$$\times B(k)^T(M(k+1) + S_{N-k-1})F(k)x(k) \triangleq A_{N-k}(k)^T x(k). \quad (4.28)$$

Substituting back to Eq. (4.27), one obtains a recursive expression for S_{N-k}:

$$S_{N-k} = \Big[F(k) - B(k)\big[B(k)^T(M(k+1) + S_{N-k-1})B(k) + N(k)\big]^{-1}$$
$$\times B(k)^T(M(k+1) + S_{N-k-1})F(k)\Big]^T (M(k+1) + S_{N-k-1})$$
$$\times\Big[F(k) - B(k)\big[B(k)^T(M(k+1) + S_{N-k-1})B(k) + N(k)\big]^{-1}$$
$$\times B(k)^T(M(k+1) + S_{N-k-1})F(k)\Big]$$
$$+ F(k)^T(M(k+1) + S_{N-k-1})B(k)$$
$$\times\big[B(k)^T(M(k+1) + S_{N-k-1})B(k) + N(k)\big]^{-1}$$
$$\times N(k)\big[B(k)^T(M(k+1) + S_{N-k-1})B(k) + N(k)\big]^{-1}$$
$$\times B(k)^T(M(k+1) + S_{N-k-1})F(k)$$
$$= \big[F(k) - B(k)A_{N-k}^T\big]^T(M(k+1) + S_{N-k-1})$$
$$\times\big[F(k) - B(k)A_{N-k}^T\big] + A_{N-k}N(k)A_{N-k}^T,$$
$$S_0(N) = 0, \quad (4.29)$$

and

$$\begin{aligned} q_{N-k} &= q_{N-k-1} + E\{w(k)^T G(k)^T (M(k+1) + S_{N-k-1}) G(k) w(k)\} \\ &= q_{N-k-1} + \mathrm{Tr}\left[(M(k+1) + S_{N-k-1}) G(k) Q(k) G(k)^T\right], \\ & \qquad q_0(N) = 0. \end{aligned} \tag{4.30}$$

Solving eq. (4.29) and (4.30) backwards and substituting into Eq. (4.24), the optimal cost for observable states is

$$I_0^*(x_0) = x_0^T S_N x_0 + \mathrm{Tr}[S_N P_0] + \sum_{k=1}^{N-1} \mathrm{Tr}\left[S_{N-k-1} G(k) Q(k) G(k)^T\right]. \tag{4.31}$$

The results are consistent with the literature [Lewis (1986)].

4.4.2 Incomplete State Information

When the information about the states is provided through noise-corrupted measurements $z(k)$, the above derivation cannot be implemented. The following device is used in order to obtain the optimal solution. Consider Eq. (4.27), where the value $u^*(k)$ of Eq. (4.28) has replaced the appropriate terms in the equation:

$$\begin{aligned} &V_{N-k}(x(k)/Z^k) \\ &\quad = \min_{u(k)} E\{(u(k) - u^*(k))^T D_{N-k}(u(k) - u^*(k)) \\ &\qquad\qquad + x(k)^T S_{N-k} x(k) + q_{N-k}/Z^k\} \\ &\quad = \min_{u(k)} E\{(u(k) - u^*(k))^T \left[B(k)^T (M(k+1) \right. \\ &\qquad\qquad \left. + S_{N-k-1}) B(k) + N(k)\right] \\ &\qquad\qquad \times (u(k) - u^*(k)) + x(k)^T S_{N-k} x(k) + q_{N-k}/Z^k\}. \end{aligned} \tag{4.32}$$

It is obvious that only the first of the three terms in Eq. (4.32) depends on $u(k)$ for minimization; for example,

$$\begin{aligned} I'(u(k)) &= E\{(u(k) - u^*(k))^T D_{N-k}(u(k) - u^*(k))/Z^k\}, \\ D_{N-k} &= B(k)^T (M(k+1) + S_{N-k-1}) B(k) + N(k). \end{aligned} \tag{4.33}$$

It is also clear from Eq. (4.28) that $u^*(k)$ is not realizable anymore, but can be decomposed into two orthogonal vectors, $\hat{u}^*(k/k) \in \{Z^k\}$, which can be estimated, and $\tilde{u}^*(k/k) \perp \{Z^k\}$, which cannot. Because the system is linear and the noise processes Gaussian one may select $\hat{u}(k/k)$ to be a linear function of the best estimate $x(k/k)$, obtained from a Kalman filter.

$$u^*(k) = \hat{u}^*(k/k) + \tilde{u}^*(k/k) = -A_{N-k}^T\hat{x}(k/k) - A_{N-k}^T\tilde{x}(k/k) \tag{4.34}$$

where the estimation part of the problem is given by:

$$\begin{aligned}
\hat{x}(k/k) &= F(k-1)\hat{x}(k-1/k-1) + \tilde{P}(k/k)H(k)^T R(k)^{-1} \\
&\quad \times [z(k) - H(k)F(k-1)\hat{x}(k-1/k-1)] \\
&\quad + \tilde{P}(k/k)\tilde{P}(k/k-1)^{-1}B(k-1)u(k-1) \\
\tilde{P}(k/k) &= \tilde{P}(k/k-1) - \tilde{P}(k/k-1)H(k)^T \\
&\quad \times \left[H(k)\tilde{P}(k/k-1)H(k)^T + R(k)\right]^{-1} H(k)\tilde{P}(k/k-1) \\
\tilde{P}(k/k-1) &= F(k-1)\tilde{P}(k-1/k-1)F(k)^T \\
&\quad + G(k-1)Q(k-1)G(k-1)^T \qquad (4.35) \\
\tilde{x}(k/k) &= \hat{x}(k/k) - x(k), \qquad \tilde{P}(k/k) = E\{\tilde{x}(k/k)\tilde{x}(k/k)^T\} \\
\hat{x}(0/0) &= E\{x(0)\}, \qquad \tilde{P}(0/0) = E\{(\hat{x}(0/0) - x_0)(\hat{x}(0/0) - x_0)^T\} = P_0.
\end{aligned}$$

Substituting eqs. (4.34) and (4.35) into eq. (4.33), one has to minimize:

$$\begin{aligned}
I'(u^*(k)) = \min_{u(k)} \Big[&(u(k) - \hat{u}^*(k/k))^T D_{N-k}(u(k) - \hat{u}^*(k/k)) \\
&+ \mathrm{Tr}\left[A_{N-k}D_{N-k}A_{N-k}^T\tilde{P}(k/k)\right]
\end{aligned}$$

where only the first term depends on $u(k)$. Therefore, for the optimal control part of the problem, the minimum occurs when:

$$\begin{aligned}
u^*(k) &= -A_{N-k}\hat{x}(k/k) \\
&= -\left[B(k)^T(M(k+1) + S_{N-k-1})B(k) + N(k)\right]^{-1} \\
&\quad \times B(k)^T(M(k+1) + S_{N-k-1})F(k)\hat{x}(k/k) \qquad (4.36)
\end{aligned}$$

Equation (4.32), reconfirms the separation theorem, that the optimal solution of the linear quadratic gaussian stochastic control problem is given by cascading a Kalman filter with a deterministic optimal control. The optimal cost is obtained from equation (4.26), by adding the cost of estimation, which is obtained from the second term of $I(u^*(k))$:

$$\begin{aligned} I_0^*(x_0) = x_0^T S_N x_0 &+ \mathrm{Tr}[S_N P_0] \\ &+ \sum_{k=1}^{N-1} \Big\{ \mathrm{Tr}\big[S_{N-k-1} G(k) Q(k) G(k)^T\big] \\ &\qquad + \mathrm{Tr}\big[A_{N-k} D_{N-k} A_{N-k}^T \tilde{P}(k/k)\big]\Big\} \end{aligned}$$

or

$$\begin{aligned} I_0^*(x_0) &= x_0^T S_N x_0 + \mathrm{Tr}[S_N P_0] \\ &+ \sum_{k=1}^{N-1} \Big\{ \mathrm{Tr}\big[S_{N-k-1} G(k) Q(k) G(k)^T\big] \\ &+ \mathrm{Tr}\big[A_{N-k}\big[B(k)^T (M(k+1) + S_{N-k-1}) B(k) + N(k)\big] \\ &\qquad \times A_{N-k}^T \tilde{P}(k/k)\big]\Big\}. \qquad (4.37) \end{aligned}$$

The following is a summary of the results of the optimal LQG discrete-time problem.

The Estimation Part

$$\begin{aligned} \hat{x}(k/k) &= F(k-1)\hat{x}(k-1/k-1) + \tilde{P}(k/k) H(k)^T R(k)^{-1} \\ &\quad \times [z(k) - H(k) F(k-1) \hat{x}(k-1/k-1)] \\ &\quad + \tilde{P}(k/k) \tilde{P}(k/k-1)^{-1} B(k-1) u^*(k-1), \\ \tilde{P}(k/k) &= \tilde{P}(k/k-1) - \tilde{P}(k/k-1) H(k)^T \\ &\quad \times \big[H(k)\tilde{P}(k/k-1)H(k)^T + R(k)\big]^{-1} H(k)\tilde{P}(k/k-1), \\ \tilde{P}(k/k-1) &= F(k-1)\tilde{P}(k-1/k-1)F(k-1)^T \\ &\quad + G(k-1)Q(k-1)G(k-1)^T, \end{aligned}$$

$$\tilde{x}(k/k) = \hat{x}(k/k) - x(k), \qquad \tilde{P}(k/k) = E\{\tilde{x}(k/k)\tilde{x}(k/k)^T\},$$

$$\hat{x}(0/0) = E\{x(0)\}, \qquad \tilde{P}(0/0) = E\{(\hat{x}(0/0) - x_0)(\hat{x}(0/0) - x_0)^T\} = P_0.$$

The Control Part

$$
\begin{aligned}
u^*(k) &= -A_{N-k}\hat{x}(k/k) \\
&= -\left[B(k)^T(M(k+1) + S_{N-k-1})B(k) + N(k)\right]^{-1} \\
&\quad \times B(k)^T(M(k+1) + S_{N-k-1})F(k)\hat{x}(k/k),
\end{aligned}
$$

$$
\begin{aligned}
S_{N-k} = \Big[&F(k) - B(k)\left[B(k)^T(M(k+1) + S_{N-k-1})B(k) + N(k)\right]^{-1} \\
&\times B(k)^T(M(k+1) + S_{N-k-1})F(k)\Big]^T (M(k+1) + S_{N-k-1}) \\
&\times\Big[F(k) - B(k)\left[B(k)^T(M(k+1) + S_{N-k-1})B(k) + N(k)\right]^{-1} \\
&\times B(k)^T(M(k+1) + S_{N-k-1})F(k)\Big] \\
&+ F(k)^T(M(k+1) + S_{N-k-1})B(k) \\
&\times \left[B(k)^T(M(k+1) + S_{N-k-1})B(k) + N(k)\right]^{-1} \\
&\times N(k)\left[B(k)^T(M(k+1) + S_{N-k-1})B(k) + N(k)\right]^{-1} \\
&\times B(k)^T(M(k+1) + S_{N-k-1})F(k),
\end{aligned}
$$

$$S_0(N) = 0.$$

AN ILLUSTRATIVE EXAMPLE In order to demonstrate the feedback solution of the stochastic optimal control problem the following example is presented. Consider the system

$$
\begin{aligned}
x(k+1) &= x(k) + u(k) + w(k), \qquad x(0) \sim N(x_0, p_0), \\
z(k) &= x(k) + v(k), \qquad w(k) \sim N(0, q), \qquad v(k) \sim N(0, r) \text{ i.i.d.}
\end{aligned}
$$

Consider also the performance cost of the process:

$$I(u) = E\left\{x(N)^2 + \sum_{i=0}^{N-1} u(i)^2\right\}$$

Solution

The Estimation Part. From Eqs. (4.35), and for $F = B = G = H = 1$, $M = 0$, $N = 1$, $P_0 = p_0$, $Q = q$, and $R = r$, one obtains

$$\hat{x}(k/k) = \hat{x}(k-1/k-1) + \tilde{P}(k/k)/r[z(k) - \hat{x}(k-1/k-1)] + \tilde{P}(k/k)\tilde{P}(k/k-1)^{-1}u(k-1),$$

$$\tilde{P}(k/k-1) = \tilde{P}(k-1/k-1) + q,$$

$$\tilde{P}(k/k) = \tilde{P}(k/k-1)\big[1 - \tilde{P}(k/k-1)/\big(\tilde{P}(k/k-1) + r\big)\big] = \big[\tilde{P}(k-1/k-1) + q\big]/\big[1 + \big(\tilde{P}(k-1/k-1) + q\big)/r\big].$$

Solving the last equation iteratively from $\tilde{P}(0/0) = p_0$,

$$\tilde{P}(k/k) = [p_0 + q]r/[r + k(p_0 + q)],$$

$$\tilde{P}(k/k)\tilde{P}(k/k-1)^{-1} = [p_0 + q]/[r + k(p_0 + q)].$$

The *optimal estimator* then becomes

$$\begin{aligned}\hat{x}(k/k) &= \hat{x}(k-1/k-1) + (p_0 + q)/[r + k(p_0 + q)] \\ &\quad \times [z(k) - \hat{x}(k-1/k-1)] \\ &\quad + [1 - (p_0 + q)/(r + k(p_0 + q))]u^*(k-1) \\ &= \hat{x}(k-1/k-1) + (p_0 + q)/[r + k(p_0 + q)] \\ &\quad \times [z(k) - \hat{x}(k-1/k-1) - u^*(k-1)] + u^*(k-1).\end{aligned}$$

The Control Part. From Eqs. (4.28) and (4.29):

$$u^*(k) = -[S_{N-k-1}/(1 + S_{N-k-1})]\hat{x}(k/k)$$

and

$$\begin{aligned}S_{N-k} &= [1 - S_{N-k-1}/(1 + S_{N-k-1})]S_{N-k-1}[1 - S_{N-k-1}/(1 + S_{N-k-1})] \\ &\quad + [S_{N-k-1}/(1 + S_{N-k-1})]^2 \\ &= S_{N-k-1}/(1 + S_{N-k-1}) = A_{N-k}^T, \qquad S_0(N) = 1.\end{aligned}$$

Solving the last equation backward,

$$S_{N-k} = 1/(N-k+1), \qquad S_{N-k-1} = 1/(N-k),$$

and the *optimal control* is

$$\begin{aligned} u^*(k) &= -\hat{x}(k/k)/(N-k+1), \\ u^*(k-1) &= -\hat{x}(k-1/k-1)/(N-k+2). \end{aligned}$$

Collecting the optimal estimator with the optimal control, one gets

$$\begin{aligned} u^*(k) = & -(1/(N-k+1)) \\ & \times\{\hat{x}(k-1/k-1) + (p_0+q)/[r+k(p_0+q)] \\ & \times[z(k) - \hat{x}(k-1/k-1) + \hat{x}(k-1/k-1)/(N-k+2)] \\ & \quad -\hat{x}(k-1/k-1)/(N-k+2)\} \\ = & -(1/(N-k+1))\{x(k-1/k-1)(N-k+1/N-k+2) \\ & +(p_0+q)/[r+k(p_0+q)][z(k) - x(k-1/k-1) \\ & \times(N-k+1/N-k+2)]\} \\ = & -(1/(N-k+2))\{\hat{x}(k-1/k-1) \\ & \qquad +(p_0+q)/[r+k(p_0+q)] \\ & \times[(1-(1/(N-k+1)))(z(k) - \hat{x}(k-1/k-1))]\}. \end{aligned}$$

The optimal cost for this example [Eq. (4.37)], is

$$\begin{aligned} I(u^*) = & \left(x_0^2 + p_0\right)/(N+1) + \sum_{k=1}^{N-1} 1/(N-k) \\ & \times\{q + (p_0+q)/[1+k(p_0+q)/r]/N-k+1\}. \end{aligned}$$

4.5 ACTIVE FEEDBACK VERSUS OPEN-LOOP CONTROL

In order to establish the global optimality of the active feedback control solution, a comparison of its cost with that of the open-loop solution is made. The system used for comparison is the LQG process, which is neutral, and therefore there is no distinction between active and passive feedback. The difference between the latter will be established later, when the entropy formulation is introduced. Because the purpose of this comparison is to demonstrate the lower cost of the feedback over the open-loop solution, an

LQG system with complete state information and fixed initial position will be assumed, avoiding the Kalman filter equations. The open-loop solution contains only a priori information on the process noise $w(k)$, and accepts, as in the deterministic case where open loop and feedback are equivalent, a controller that depends on the initial state x_0:

$$u^*(x_0, k)_{\mathrm{OL}} = -A_{N-k}^T \Phi(k) x_0, \tag{4.38}$$

where A_{N-k}^T is the deterministic optimal control gain of Eq. (4.28), and $\Phi(k)$ is the state transition matrix associated with the deterministic closed-loop system:

$$\Phi(k) \triangleq \prod_{i=0}^{k-1} \left[F(i) - B(i) A_{N-i}^T\right]. \tag{4.39}$$

The stochastic optimal feedback control is given by Eq. (4.28):

$$u^*(x(k), k)_{\mathrm{FB}} = -A_{N-k}^T x(k). \tag{4.40}$$

Comparing $u^*(x_0, k)_{\mathrm{OL}}$ with $u^*(x(k), k)_{\mathrm{FB}}$, it is easily verified that

$$u^*(x_0, k)_{\mathrm{OL}} = u^*(x(k), k)_{\mathrm{FB}} + A_{N-k}^T \sum_{i=0}^{k-1} \Psi(k-i-1) G(i) w(i), \tag{4.41}$$

where $\Psi(\cdot)$ is the state transition matrix associated with the open-loop system:

$$\Psi(k-i-1) = \prod_{j=0}^{k-i-1} F(j). \tag{4.42}$$

Substituting $u^*(x_0, k)_{\mathrm{OL}}$ and $u^*(x(k), k)_{\mathrm{FB}}$ in the cost function Eq. (4.24) to be minimized, respectively, and comparing after some algebra:

$$I(u_{\mathrm{OL}}^*) - I(u_{\mathrm{FB}}^*) = \sum_{i=0}^{N-1} \mathrm{Tr}\Bigg[G(i) Q(i) G(i)^T \sum_{k=i+1}^{N-1} \Psi(k-i-1) \times A_{N-k} N(k) A_{N-k}^T \Psi(k-i-1)\Bigg] > 0. \tag{4.43}$$

This proves that the cost of the stochastic open-loop optimal solution is higher than the feedback optimal even in the absence of measurement noise.

The behavior of the stochastic feedback optimal solution of the LQG problem presents certain differences from the deterministic case. This is in addition to the differences among the open-loop, open-loop feedback, and closed-loop solutions that arise by the way information about the system uncertainties are handled.

For instance, although the controller is deterministic, the cost that it minimizes is only the expected value and not necessarily the true value of the performance of the system, which can be measured only after the execution of the operation. The same thing can be said about the stability of the system, if it is assessed by the cost functional. These topics will be discussed later in conjunction with the continuous-time case.

4.6 LINEAR-QUADRATIC-GAUSSIAN OPTIMAL-CONTROL PROBLEM: THE CONTINUOUS-TIME CASE

Consider the linear time-varying continuous-time system:

$$\begin{aligned} dx/dt &= F(t)x(t) + B(t)u(t) + G(t)w(t), \qquad x(0), \\ z(t) &= H(t)x(t) + v(t), \end{aligned} \tag{4.44}$$

where $x(t)$ is the state of the system at time t, $z(t)$ is the current m-dimensional measurement vector, $u(t)$ is the r-dimensional control vector, and $F(t)$, $B(t)$, $G(t)$, and $H(t)$ are $n \times n$, $n \times r$, $n \times r$, and $m \times n$ matrices of known coefficients, respectively. The stochastic variable $x(0)$ is the n-dimensional Gaussian initial state, $w(t)$ is the r-dimensional process noise, and $v(t)$ is the m-dimensional measurement noise, for all $t \in [0, \infty]$, satisfying

$$\begin{aligned} E\{x(0)\} &= x_0, \qquad E\{(x(0) - x_0)(x(0) - x_0)^T\} = P_0, \\ E\{w(t)\} &= 0, \qquad E\{w(t)w(\tau)^T\} = Q(t)\,\delta(t - \tau), \\ E\{v(t)\} &= 0, \qquad E\{v(t)v(\tau)^T\} = R(t)\,\delta(t - \tau). \\ E\{x(0)w(t)^T\} &= E\{x(0)v(t)^T\} = E\{w(t)v(\tau)^T\} = 0. \end{aligned} \tag{4.45}$$

A cost functional containing all the predefined desired specifications about the system's performance is also defined

$$\begin{aligned} I(u) &= E\{J(u)\} \\ &= \tfrac{1}{2}E\left\{x(T)^T\Phi x(T) + \int_0^T \left[x(\tau)^T M(\tau)x(\tau) + u(\tau)^T N(\tau)u(\tau)\right] d\tau\right\}. \end{aligned} \tag{4.46}$$

The problem is to find a u^*:

$$I(u^*) = I^* = \min_u I(u)$$

4.6.1 Complete State Information

In order to obtain the optimal solution of the continuous-time LQG problem, the Hamilton–Jacobi–Bellman equation (4.21) will be used, where a quadratic form for the $V(t/X^t)$ function is assumed. This will convert the partial differential Eq. (4.21) into an ordinary differential equation:

$$V(t/X^t) = \tfrac{1}{2}x(t)^T S(t)x(t) + q(t), \qquad S(T) = \Phi, \qquad q(T) = 0. \quad (4.47)$$

By differentiation one obtains

$$\begin{aligned}
\partial V/\partial t &= \tfrac{1}{2}x(t)^T\, dS(t)/dt\, x(t) + q(t),\\
\partial V/\partial x &= S(t)x(t),\\
\partial^2 V/\partial x^2 &= S(t).
\end{aligned}$$

Substitution into Eq. (4.21), yields

$$\begin{aligned}
&\tfrac{1}{2}x^T\, dS/dt\, x + dq/dt + \min_u \left[\tfrac{1}{2}(x^T Mx + u^T Nu) + x^T SFx + x^T SBu\right]\\
&\quad + \tfrac{1}{2}\,\mathrm{Tr}\left[SGQG^T\right] = 0. \qquad (4.48)
\end{aligned}$$

Because Eq. (4.48) must hold for all $x(t)$'s each term of the quadratic must be equal to zero; for example,

$$\tfrac{1}{2}x^T\, dS/dt\, x + \min_u \left[\tfrac{1}{2}(x^T Mx + u^T Nu) + x^T SFx + x^T SBu\right] = 0, \qquad S(T) = \Phi,$$

$$dq(t)/dt = -\tfrac{1}{2}\,\mathrm{Tr}\left[S(t)GQG^T\right], \qquad q(T) = 0.$$

Minimization of the first equation, and substitution back, yields

$$\begin{aligned}
u^*(x,t) &= -N(t)^{-1}B^T S(t)x(t),\\
x(t)^T\left[dS/dt + M + SF + F^T S - SBN^{-1}B^T S\right]x(t) &= 0, \qquad \forall\, x(t).
\end{aligned} \qquad (4.49)$$

Summarizing the results of the optimization for the continuous-time LQG problem with complete information on the states, for example, $z(t) \equiv x(t)$:

$$\begin{aligned}
u^*(x,t) &= -N(t)^{-1}B^T S(t)x(t),\\
dS/dt + M + SF + F^T S - SBN^{-1}B^T S &= 0, \qquad S(T) = \Phi,\\
q(t) &= \tfrac{1}{2}\int_t^T \mathrm{Tr}\left[S(\tau)G(\tau)Q(\tau)G(\tau)^T\right] d\tau.
\end{aligned} \qquad (4.50)$$

The second of Eqs. (4.50) is a matrix Riccati equation very similar to the one encountered in the Kalman filter [Eq. (4.21)], and it may be solved in a similar way [Eqs. (4.27)–(4.35)], but with boundary conditions at time T. The cost of performance is given by

$$I(u^*(x_0, 0)) = \tfrac{1}{2}x_0^T S(0) x_0 + \tfrac{1}{2}\int_0^T \mathrm{Tr}\left[S(\tau)G(\tau)Q(\tau)G(\tau)^T\right] d\tau. \quad (4.51)$$

4.6.2 Incomplete State Information

A slightly different approach than the one used in the discrete-time case will be followed to obtain the continuous-time LQG optimal control with incomplete state information. The method resembles Kalman's approach to obtain the feedback optimal control for the deterministic case. In this problem the information about the states is available through the measurement equation of Eq. (4.44), and the m-dimensional vector $z(t)$. Consider now the performance cost function, given by eq. (4.46) conditioned on the measurements Z^t, and starting at an arbitrary time t:

$$\begin{aligned} &I(u^*(x, t)) \\ &\quad = E\{V(x(t)/Z^t)\} = \tfrac{1}{2}E\{x(t)^T S(t) x(t)/Z^t\} + \tfrac{1}{2}\int_t^T \mathrm{Tr}[S(\tau)GQG^T]\, d\tau \\ &\quad = \tfrac{1}{2}E\left\{x(T)^T \Phi x(T) + \int_t^T [x(\tau)^T M(\tau) x(\tau) + u^*(\tau)^T N(\tau) u^*(\tau)]\, d\tau/Z^t\right\}, \end{aligned} \quad (4.52)$$

$$E\{V(x(T)/Z^T)\} - E\{V(x(t)/Z^t)\} = \int_t^T d\left[E\{V(x(\tau)/Z^\tau)\}/d\tau\right] d\tau, \quad (4.53)$$

and for stochastic differentiation of $E\{V(x(t)/Z^t)$;

$$\begin{aligned} &dE\{V(x(t)/Z^t)\}/dt \\ &\quad = \tfrac{1}{2} d\left[E\{x(t)^T S(t) x(t)/Z^t\} + \int_t^T \mathrm{Tr}[S(\tau)GQG^T]\, d\tau\right]\Big/ dt \\ &\quad = \tfrac{1}{2}E\{x^T\, dS/dt\, x + 2x^T SFx + 2x^T SBu^* + 2x^T SGw/Z^t\} \\ &\qquad - \tfrac{1}{2}\,\mathrm{Tr}\, S(t) GQG^T \\ &\quad = \tfrac{1}{2}E\{x^T\, dS/dt\, x + x^T SFx + x^T F^T Sx + x^T SBu^* + u^* B^T Sx/Z^t\}. \end{aligned} \quad (4.54)$$

Now subtract both sides of Eq. (4.53) from Eq. (4.52), substituting Eq. (4.54), and collecting terms:

$$\begin{aligned}
I(u^*(x,t)) &= \tfrac{1}{2}E\{x(T)^T \Phi x(T)/Z^t\} \\
&\quad + \tfrac{1}{2}E\left\{\int_t^T \left[x(\tau)^T M(\tau)x(\tau) + u^*(\tau)^T N(\tau)u^*(\tau)\right] d\tau/Z^t\right\} \\
&\quad - E\{V(x(T)/Z^T)\} + E\{V(x(t)/Z^t)\} \\
&\quad + \tfrac{1}{2}E\left\{\int_t^T \left[x^T\, dS/dt\, x + x^T SFx + x^T F^T Sx + x^T SBu^* + u^* B^T Sx\right] d\tau/Z^t\right\} \\
&= \tfrac{1}{2}E\{x(t)^T S(t)x(t)/Z^t\} + \tfrac{1}{2}\int_t^T \mathrm{Tr}\left[S(\tau)GQG^T\right] d\tau \\
&\quad + \tfrac{1}{2}E\left\{\int_t^T \left[x(\tau)^T M(\tau)x(\tau) + u^*(\tau)^T N(\tau)u^*(\tau)\right] d\tau/Z^t\right\} \\
&\quad + \tfrac{1}{2}E\left\{\int_t^T \left[x(\tau)^T\left[dS/dt + SF + F^T S\right]x(\tau)\right.\right. \\
&\qquad\qquad \left.\left. + x^T SBu^* + u^* B^T Sx\right] d\tau/Z^t\right\}.
\end{aligned}$$

Let $t = 0$, the start of the process,

$$\begin{aligned}
I(u^*(x,0)) &= \tfrac{1}{2}E\{x(0)^T S(0)x(0)/Z^t\} + \tfrac{1}{2}\int_0^T \mathrm{Tr}\left[S(\tau)GQG^T\right] d\tau \\
&\quad + \tfrac{1}{2}E\left\{\int_0^T \left[x(\tau)^T\left[M + dS/dt + SF + F^T S\right]x(\tau) + x^T SBu^*\right.\right. \\
&\qquad \left.\left. + u^* B^T Sx + u^*(\tau)^T N(\tau)u^*(\tau)\right] d\tau/Z^t\right\}. \quad (4.56)
\end{aligned}$$

Considering Eq. (4.49),

$$M + dS/dt + SF + F^T S = SBN^{-1}B^T S,$$

and substituting back into Eq. (4.56) yields

$$\begin{aligned}
I(u^*(x,0)) &= \tfrac{1}{2}E\{x(0)^T S(0)x(0)\} + \tfrac{1}{2}\int_0^T \mathrm{Tr}\left[S(\tau)GQG^T\right] d\tau \\
&\quad + \tfrac{1}{2}\min_u E\left\{\int_t^T \| N^{-1}B^T Sx(\tau) + u(\tau)\|_N^2\, d\tau/Z^t\right\}. \quad (4.57)
\end{aligned}$$

Minimization, for complete state information, yields

$$u^*(x,t) = -N(t)^{-1}B^T S(t)x(t).$$

Because $x(t)$ is not available for measurement, one may decompose the state vector into two orthogonal vectors, $\hat{x}(t/t) \perp \tilde{x}(t/t)$;

$$x(t) = \hat{x}(t/t) + \tilde{x}(t/t),$$

where $\hat{x}(t/t)$ is the output of a Kalman–Bucy filter, with the optimal-control input (observe the gain and compare with the discrete-time case);

$$\begin{aligned} d\hat{x}(t/t)/dt &= \left[F(t) - B(t)N(t)^{-1}B^T(t)S(t)\right]\hat{x}(t/t) \\ &\quad + K(t)[z(t) - H(t)\hat{x}(t/t)] \\ K(t) &= \tilde{P}(t/t)H(t)^T R(t)^{-1}, \qquad \hat{x}(0/0) = x_0, \qquad \tilde{P}(0/0) = P_0; \end{aligned} \tag{4.58}$$

$$\begin{aligned} d\tilde{P}(t/t)/dt &= F(t)\tilde{P}(t/t) + \tilde{P}(t/t)F(t)^T \\ &\quad - K(t)H(t)\tilde{P}(t/t) + G(t)Q(t)G(t)^T. \end{aligned}$$

Substituting the decomposed state into Eq. (4.57),

$$\begin{aligned} I(u^*(x,0)) &= \tfrac{1}{2}E\{x(0)^T S(0)x(0)/Z^t\} + \tfrac{1}{2}\int_0^T \mathrm{Tr}\left[S(\tau)GQG^T\right] d\tau \\ &\quad + \tfrac{1}{2}\min_u E\left\{\int_t^T \| N^{-1}B^T S\hat{x}(\tau/\tau) + u(\tau)\|_N^2 \, d\tau/Z^t\right\} \\ &\quad + \tfrac{1}{2}E\left\{\int_t^T \| N^{-1}B^T S\tilde{x}(\tau/\tau)\|_N^2 \, d\tau/Z^t\right\}. \end{aligned} \tag{4.59}$$

In Eq. (4.59), only the third term can be affected by minimization, and the other ones are constant. This yields an optimal control:

$$\begin{aligned} u^*(x,t) &= -N(t)^{-1}B^T S(t)\hat{x}(t/t), \\ dS/dt + M + SF + F^T S - SBN^{-1}B^T S &= 0, \qquad S(T) = \Phi, \\ q(t) &= \tfrac{1}{2}\int_t^T \mathrm{Tr}\left[S(\tau)G(\tau)Q(\tau)G(\tau)^T\right] d\tau, \end{aligned} \tag{4.60}$$

with $\hat{x}(t/t)$ the output of the Kalman filter;

$$d\hat{x}(t/t)/dt = \left[F(t) - B(t)N(t)^{-1}B^T(t)S(t)\right]\hat{x}(t/t) + K(t)[z(t) - H(t)\hat{x}(t/t)],$$

$$K(t) = \tilde{P}(t/t)H(t)^T R(t)^{-1}, \qquad \hat{x}(0/0) = x_0, \qquad \tilde{P}(0/0) = P_0, \tag{4.61}$$

$$d\tilde{P}(t/t)/dt = F(t)\tilde{P}(t/t) + \tilde{P}(t/t)F(t)^T - K(t)H(t)\tilde{P}(t/t) + G(t)Q(t)G(t)^T.$$

The above derivation establishes the validity of the *separation theorem*, where the stochastic optimal control is generated by cascading a deterministic optimal controller with a Kalman filter. This is a very exciting result, because of its simplicity. Unfortunately, it has created many incorrect interpretations when the system was not linear quadratic Gaussian. The *performance cost* of this operation is given from Eq. (4.59):

$$I(u^*(x,0)) = \tfrac{1}{2}E\{x(0)^T S(0)x(0)/Z^t\} + \tfrac{1}{2}\int_0^T \mathrm{Tr}[S(\tau)GQG^T]\,d\tau + \tfrac{1}{2}\int_0^T \mathrm{Tr}\left[\tilde{P}(\tau/\tau)S(\tau)B(\tau)B(\tau)N^{-1}B(\tau)^T S(\tau)\right]d\tau.$$

Decomposing the first term,

$$\tfrac{1}{2}E\{x(0)^T S(0)x(0)/Z^t\} = \tfrac{1}{2}\mathrm{Tr}\left[S(0)\mathrm{cov}\{x(0)x(0)^T\}\right]$$

$$\mathrm{cov}\{x(0)x(0)^T\} = x_0 x_0^T + P_0.$$

The *performance cost* becomes

$$I(u^*(x,0)) = \tfrac{1}{2}x_0^T S(0)x_0 + \tfrac{1}{2}\mathrm{Tr}[S(0)P_0] + \tfrac{1}{2}\int_0^T \mathrm{Tr}[S(\tau)GQG^T]\,d\tau + \tfrac{1}{2}\int_0^T \mathrm{Tr}\left[\tilde{P}(\tau/\tau)S(\tau)B(\tau)B(\tau)N^{-1}B(\tau)^T S(\tau)\right]d\tau,$$

where

$$\tfrac{1}{2}x_0^T S(0)x_0 = \text{deterministic cost of control}$$

$$\tfrac{1}{2}\mathrm{Tr}[S(0)P_0] = \text{cost of uncertainty of initial position}$$

$$\tfrac{1}{2}\int_0^T \mathrm{Tr}[S(\tau)GQG^T]\,d\tau = \text{cost due to process noise}$$

$$\tfrac{1}{2}\int_t^T \mathrm{Tr}\left[\tilde{P}(\tau/\tau)S(\tau)B(\tau)B(\tau)N^{-1}B(\tau)^T S(\tau)\right]d\tau = \text{cost due to measurement noise}.$$

AN ILLUSTRATIVE EXAMPLE The following academic example illustrates the method. Consider the system

$$dx(t)/dt = -\tfrac{1}{2}x(t) + u(t) + w(t),$$
$$z(t) = x(t) + v(t).$$

The performance cost is given:

$$J(u) = \tfrac{1}{2}E\left\{2x(T)^2 + \int_0^T \left[2x(\tau)^2 + u(\tau)^2\right] d\tau\right\}.$$

The initial position is fixed:

$$x(0) = 10,$$

and the process and measurement uncertainties, respectively, are

$$w(t) \sim N(0, \tfrac{1}{4}), \qquad v(t) \sim N(0, 2)$$

The solution is formulated by cascading a Kalman–Bucy filter with a deterministic optimal controller.

The Kalman–Bucy Filter

$$\begin{aligned} d\hat{x}/dt &= -\tfrac{1}{2}\hat{x}(t/t) + K(t)[z(t) - \hat{x}(t/t)] + u^*(x) \\ &= -\tfrac{1}{2}\hat{x}(t/t) + 4\tilde{P}(t)[z(t) - \hat{x}(t/t)] - C(t)\hat{x}(t/t), \\ K(t) &= 4\tilde{P}(t), \qquad \hat{x}(0/0) = 10, \\ d\tilde{P}(t)/dt &= -\tilde{P}(t) - 4\tilde{P}(t)^2 + 2, \qquad \tilde{P}(0) = 0. \end{aligned}$$

Solving the matrix Riccati equation yields

$$\tilde{P}(t) = -\tfrac{1}{8} - \sqrt{33}/8 \tanh\left[\sqrt{33}/8t + \tanh^{-1} 1/\sqrt{33}\right],$$

and the Kalman filter gain is

$$K(t) = -\tfrac{1}{2} - \sqrt{33}/2 \tanh\left[\sqrt{33}/2t + \tanh^{-1} 1/\sqrt{33}\right].$$

The Optimal Control

$$u^*(x) = -C(t)x(t/t),$$
$$C(t) = s(t),$$
$$ds/dt + s(t) + s(t)^2 - 2 = 0, \qquad T = 1, \qquad s(T) = 2.$$

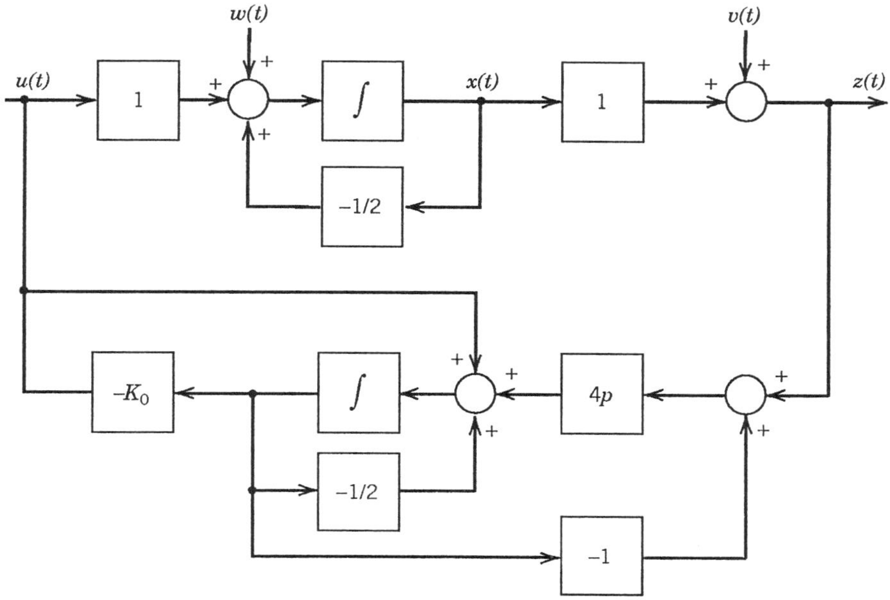

Fig. 4.2 Continuous-time stochastic optimal LQG control

Solving the control matrix Riccati equation, which is dual to the estimation equivalent,

$$s(t) = [4\exp\{-3(t-1)\} + 2]/[4\exp\{-3(t-1)\} - 1],$$

and the optimal controller is

$$\begin{aligned} u^*(x) &= -s(t)x(t/t) \\ &= -[4\exp\{-3(t-1)\} + 2]/[4\exp\{-3(t-1)\} - 1]\hat{x}(t/t). \end{aligned}$$

The cost of performance for this example is

$$I(u^*) = 5150 + \tfrac{1}{8}\int_0^1 s(\tau)\left[1 + 4\tilde{P}(\tau)\right] d\tau.$$

The results are depicted in Figure 4.2.

4.6.3 Some Remarks

It has been mentioned several times that the results of optimizing a stochastic LQG system via the separation theorem were too good to be generalized. A verification came later by Fel'dbaum's *dual control* formulation. A formal definition of the separation property and the weaker certainty equivalence property are given for reference.

The separation property in stochastic control is that the estimation algorithm does not depend on the optimal-control function $\Psi[\cdot]$ and vice versa; for example,

$$\Delta u(x,t) = \Psi[\Delta x(t/t)].$$

On the other hand, the *certainty equivalence property* requires that the stochastic optimal control function $\Psi[\cdot]$ be the same as the deterministic optimal control function, which is not always true for neutral systems. An example is the case of a linear Gaussian system with cost function

$$J(u) = E\left\{\mu \exp\left[\tfrac{1}{2}\mu\left[\phi(x(T)) + \int_0^T L\,d\tau\right]\right]\right\},$$

where the Kalman–Bucy filter gain does not depend on the control function, but the control function depends on the covariance of the estimator. The separation theorem is not valid here. The *Dual Control* problem, which was investigated by Fel'dbaum (1965), presents an interesting topic of investigation. Due to the dependence of the future estimation on the control actions of the future, it was shown that an analytic solution of the problem is practically impossible, except for some special cases.

A further discussion on the Dual Control problem will be postponed until a later chapter. Its performance cost, representing the cost of the active feedback control solution, will be evaluated and compared to other ones, with the help of entropy functions.

4.7 ACTIVE FEEDBACK VERSUS OPEN-LOOP CONTROL

As in the discrete-time stochastic control case the global optimality of the active feedback control will be demonstrated. An LQG system with complete state information, which was shown to be neutral, will be used for the demonstration [Wonham (1970)]. This comparison will demonstrate the lower cost of the feedback over the open-loop solution, leaving the active feedback case comparison to the LQG for a later discussion. Like in the discrete-time case, the open-loop solution contains only a priori information on the process noise $w(t)$, and accepts a controller that depends on the initial state x_0:

$$u^*(x_0,t)_{\mathrm{OL}} = -N(t)^{-1}B(t)^T S(t)\Phi(t,0)x_0 \tag{4.62}$$

where $\Phi(t,0)$ is the state transition matrix associated with the deterministic closed-loop system:

$$dx/dt = \left[F(t) - B(t)N(t)^{-1}B(t)^T S(t)\right]x(t), \qquad x(0) = x_0.$$

The stochastic optimal feedback control is given by Eq. (4.50):

$$u^*(x,t)_{\mathrm{FB}} = -N(t)^{-1}B(t)^T S(t)x(t). \tag{4.63}$$

Comparing $u^*(x_0,t)_{\mathrm{OL}}$ with $u^*(x,t)_{\mathrm{FB}}$, it is easily verified that

$$u^*(x_0,t)_{\mathrm{OL}} = u^*(x,t)_{\mathrm{FB}} + N(t)^{-1}B(t)^T S(t)\int_0^t \Psi(t-\tau)G(\tau)w(\tau)\,d\tau, \tag{4.64}$$

where $\Psi(\cdot)$ is the state transition matrix associated with the open-loop system:

$$dx/dt = F(t)x(t).$$

Substituting $u^*(x_0,t)_{\mathrm{OL}}$ and $u^*(x,t)_{\mathrm{FB}}$ in the cost function Eq. (4.46) to be minimized, respectively, and comparing after some algebra:

$$I(u^*_{\mathrm{OL}}) - I(u^*_{\mathrm{FB}}) = \int_0^T \mathrm{Tr}\left[G(\tau)A(\tau)G(\tau)^T Q(\tau)\right]d\tau > 0 \tag{4.65}$$

where the quantity $A(t)$ is defined by

$$A(t) = \int_t^T \Psi(s,t)^T S(s)B(s)N(s)^{-1}B(S)^T S(s)\Psi(s,t)\,ds. \tag{4.66}$$

From Eq. (4.65) it is obvious that $I(u^*_{\mathrm{OL}}) > I(u^*_{\mathrm{FB}})$, which establishes the claim that the feedback solution is optimal vis-à-vis the open-loop solution.

4.8 STABILITY CONSIDERATIONS

Stability is a problem related to the convergence of the states of the stochastic system to a finite value, yielding a bounded value of the cost function, as $t \to \infty$. Optimality does not always guarantee stability. On the other hand, there are different types of stochastic convergence (e.g., convergence in the mean, with probability 1, in the mean-square sense, etc.) which do not always guarantee boundedness for all types of process noise, especially when they are not bounded, like the Gaussian. In order to study the behavior of a stochastic control system as $t \to \infty$, complete state information is assumed, because the measurement equation does not affect stability, and the performance cost is modified in order to possess a finite value at infinity. Here, the continuous-time case will be considered. Generalization to the discrete-time case is straightforward.

System

$$dx/dt = Fx(t) + Bu(t) + Gw(t), \qquad x(0) \sim N(x_0, P_0), \qquad w(t) \sim N(0, Q).$$

Performance Cost

$$I(u) = \tfrac{1}{2} \lim_{T\to\infty} E\left\{(1/T)\int_0^T \left[x(\tau)^T Mx(\tau) + u(\tau)^T Nu(\tau)\right] d\tau\right\}. \tag{4.67}$$

Observe that all the coefficients are assumed to be constant, and that the cost function has been replaced by a limiting case.

4.8.1 Deterministic Case: $w(t) \equiv 0$

If the asterisk denotes the optimal values, the optimal control is

$$\begin{aligned} u^*(x) &= -Ax(t) = -N^{-1}B^T Sx(t), \\ dS/dt &= -F^T S - SF - M + SBN^{-1}B^T S, \qquad S(T) = 0, \\ I(u^*) &= \tfrac{1}{2}E\{x_0^T S(0) x_0\}. \end{aligned} \tag{4.68}$$

Existence of the solution over $[0, T < \infty]$ assures existence of $S(t)$ in the matrix Riccati equation, as this is analytic in t and S. Therefore, with bounded M and $S(T) = 0$, $S(t)$ is bounded over $[0, T]$. With $M > 0$, $S(0)$ is also bounded, because $I^*(T_2) \geq I^*(T_1)$, $T_2 \geq T_1$ $(w \equiv 0)$, and $S(T) = -M$. Therefore the control gain $A(0) < \infty$ is bounded. As $T \to \infty$, the optimal deterministic cost is

$$I(u^*) = \tfrac{1}{2}\int_0^\infty x^*(\tau)^T[M + SBN^{-1}B^T S]x^*(\tau)\, d\tau. \tag{4.69}$$

Kalman's conditions for asymptotic stability of the deterministic optimal-control problem [Kalman (1963)] are that the closed-loop system be completely controllable, the states be completely observable, (CO) $M \geq 0$, and $N > 0$. Because the states are directly measured, they are CO, and as long as $M > 0$, and the system is completely controllable (CC), $w \equiv 0$, $[M + SBN^{-1}B^T S] > 0$, and $I(u^*) < \infty$, is bounded. In other words, there exists a Lyapunov function $V(x)$;

$$V(x) = x^T S(0) x > 0,$$

where $S(0) > 0$, is the solution of the steady-state equation,

$$0 = -F^T S(0) - S(0)F - M + S(0)BN^{-1}B^T S(0), \qquad A = -N^{-1}B^T S(0).$$

The derivative of the Lyapunov function $V(x)$ along the trajectories is

$$\begin{aligned} dV(x)/dt &= \partial V/\partial x\, dx(t)/dt = 2x^T S(0)[Fx - Bu^*] \\ &= 2x^T\left[S(0)F - BN^{-1}B^T S(0)\right]x = -x^T Mx < 0, \end{aligned}$$

which means that $x(t)^*$ is globally asymptotically stable; for example,

$$\lim_{t\to\infty} x(t)^* = 0.$$

4.8.2 Stochastic Case

In the stochastic case the existence of the solution of the same matrix Riccati equation as in the previous case guarantees the boundedness of $S(0)$, as $T \to \infty$. Assuming uniform stochastic controllability and observability it suffices to show that the optimal cost is bounded for stochastic stability in the mean:

$$\begin{aligned} I(u^*) &= \lim_{T\to\infty} (1/2T) E\{x(0)^T S(0) x(0)\} \\ &\quad + \int_0^T \operatorname{Tr}\left[S(0)GQG^T\right] d\tau + \int_0^T \operatorname{Tr}\left[ANA^T P_0\right] d\tau \\ &= \lim_{T\to\infty} (1/2T)\{x_0^T S(0) x_0 + \operatorname{Tr}[S(0) P_0 \\ &\qquad + T\left[\operatorname{Tr}(S(0)GQG^T) + \operatorname{Tr}(ANA^T P_0)\right]\} \\ &= \tfrac{1}{2} \operatorname{Tr}\left[S(0)GQG^T + ANA^T P_0\right] < \infty. \end{aligned}$$

However, stability of the average performance is not the kind of stability that engineers require for the design of a system. The study of more elaborate methods, using no escape time, are too complicated to handle and are not covered in this text [Dynkin (1953)].

PROBLEMS

4.1 Define the following cases of stochastic optimal control:

- **(a)** Open-loop optimal
- **(b)** Open-loop feedback optimal
- **(c)** Neutral feedback optimal
- **(d)** Active (dual) feedback optimal

Give a mathematical description of each definition. How do these cases rank in decreasing order of cost?

4.2 For the system

$$x(k+1) = x(k) + u(k) + w(k),$$
$$z(k) = x(k) + v(k),$$

where $x(k)$ and $z(k)$ are scalars and $x(0) \sim N(x_0, P_0)$, $w(k) > N(0, q^2)$, $v(k) \sim N(0, r^2)$, and the cost function

$$J(u) = E\left\{\sum_{i=0}^{n-1}\left[x(i+1)^2 + u(i)^2\right]\right\},$$

derive the active feedback optimal control.

4.3 In the stochastic optimal-control problem

$$x(k+1) = Fx(k) + Bu(k) + Gw(k), \qquad x(0) \sim N(x_0, P_0)$$
$$z(k) = h^T x(k) + v(k), \qquad w(k) \sim N(0 \cdot q^2), v(k) \sim N(0, r^2)$$
$$J(u) = E\left\{\sum_{i=0}^{n-1}\left[x(i+1)^T Mx(i+1) + u(i)^T u(i)\right]\right\},$$

the optimal cost at the state k may be written as

$$V_{N-k}(x(k)/Z^k) = \min_{u(k)} E\{(u(k) - u(k)^*)^T D_{N-k}(u(k) - u(k)^*)$$
$$+ x(k)^T S_{N-k} x(k) + q_{N-k}/Z^k\},$$

where $u^*(k) = -A_{N-k}^T x(k)$, the optimal deterministic control with complete state information. Because the states are available only through the measurements $Z^{kT} = \{z(1) \cdots z(k)\}$, prove that the stochastic optimal control with incomplete state information is

$$u(k)^* = -A_{N-k}^T \hat{x}(k/k),$$

where $x(k/k)$ is the best estimate of the states.

4.4 Given the linear continuous-time stochastic system with constant coefficients, and fixed initial conditions x_0:

$$dx/dt = Fx(t) + Bu(t) + Gw(t),$$

where $w(t) \sim N(0, Q)$ is the process noise, and the performance index;

$$I(u) = \tfrac{1}{2} E \int_0^T (x^T Mx + u^T Nu)\, dt.$$

Assuming the general Hamilton–Jacobi–Bellman equation

$$\partial V/\partial t + \min_u \left[E\{L(x,u,t) + \partial V^T/\partial x f(x,u) + \tfrac{1}{2}\operatorname{Tr}(\partial^2 V/\partial x^2\, GQG^T)\}\right] = 0,$$

where

$$V(x,t/x) = \min_u \left[E\left\{ \int_t^T L(x,u,\tau)\, d\tau / x(t) \right\} \right].$$

Derive explicitly the solution of the optimal-control problem and obtain the optimal cost function $I(u)^* = \min_u I(u)$ as a function of the initial conditions x_0. Explain the source of each of the terms of the cost function.

4.5 Given the continuous-time stochastic system

$$\begin{aligned} dx_1/dt &= x_2, \\ dx_2/dt &= x_3, \\ dx_3 &= 0.5x_1 - 1.5x_2 + w(t) + u(t), \\ z(t) &= x_1 + v(t), \end{aligned}$$

where $w(t) \sim N(0, 0.1)$

$$v(t) \sim N(0, 0.1); \qquad x_0 \sim N\begin{bmatrix} 0 & 900 \\ 0, & 040 \\ 0 & 004 \end{bmatrix}.$$

Given also the performance cost

$$I(u) = \lim_{T\to\infty} E\left\{ (1/T) \int_0^T \left(x_2^2 + x_3^2 + 0.5u^2 \right) dt \right\},$$

derive the equations of the Kalman–Bucy filter for the estimation part, and the optimal feedback controller for the above system, and draw a block diagram to indicate their implementation.

REFERENCES

Aoki, M., *Introduction to Stochastic Systems*, Academic Press, New York, 1967.

Åström, K. J., *Introduction to Stochastic Control Theory*, Academic Press, New York, 1970.

Bellman, R., *Dynamic Programming*, Princeton University Press, Princeton, NJ, 1956.

Bellman, R., *Adaptive Control Processes, A Guided Tour*, Princeton University Press, Princeton, NJ, 1961.

Doob, J. L., *Stochastic Processes*, Wiley, New York, 1953.

Dynkin, E. B., *Markov Processes*, Academic Press, New York, 1953.

Fel'dbaum, A., *Optimal Control Systems*, Academic Press, New York, 1965.

Itô, K., *On Stochastic Differential Equations*, *Mem. Am. Math. Soc.*, Vol. 4 (1951).

Kailath, T., and Frost, D., "An Innovations Approach to Least Squares Estimation, Parts I, II, III, IV," *IEEE Trans. Autom. Control*, Vol. AC-13, No. 6, pp. 655–660 (1968).

Kalman, R. E., and Bucy, R., "New Results in Linear Filtering and Prediction Theory," *ASME Trans. J. Basic Eng*. pp. 83–95 (1961).

Kwakernaak, H., and Sivan, R., *Linear Optimal Control Systems*, Wiley, New York, 1972.

Kushner, H., *Introduction to Stochastic Control*, Holt Reinhart and Winston, New York, 1971.

Lee, R. C. K., *Optimal Identification, Estimation, and Control*, MIT Press, Cambridge, MA, 1964.

Lewis, F. L., *Optimal Estimation: Introduction to Stochastic Control*, Wiley-Interscience, New York, 1986.

McGarty, T. P., *Stochastic Systems and State Estimation*, Wiley-Interscience, New York, 1974.

Meditch, J., *Stochastic Linear Estimation and Control*, McGraw-Hill, New York, 1969.

Sage, A. P., *Optimal Systems Control*, Prentice-Hall, New York, 1968.

Sage, A. P., and White, C. C., *Optimum Systems Control*, Prentice-Hall, Englewood Cliffs, NJ, 1977.

Saridis, G. N., *Self-Organizing Controls of Stochastic Systems*, Marcel Dekker, New York, 1977.

Tou, J., and Joseph, P., "On Linear Control Theory," *AIEE Trans*., Vol. 80, No. 19 (1961).

Wonham, W. M., "Random Differential Equations in Control Theory," in A. T. Bharucha-Reid (Ed.), *Probabilistic Methods in Applied Mathematics*, Academic Press, New York, 1970.

5

APPROXIMATION THEORY FOR NONLINEAR STOCHASTIC CONTROL SYSTEMS

5.1 INTRODUCTION

The problem of controlling a stochastic, dynamic system so that its behavior is optimal with respect to a performance index has received considerable attention over the past two decades. From an optimality point of view, it is desirable to obtain a feedback solution to the optimal nonlinear control problem. In situations of linear stochastic systems with additive white Gaussian noise and quadratic performance indices (so-called LQG problems), the separation principle (or the certainty equivalence principle) is directly applicable, and the optimal-control theory is well established [Aoki (1967,) Wonham (1970), Kwakernaak and Sivan (1972), Sage and White (1977)].

However, due to the mathematical difficulties dealing with stochastic processes, only fragmentary results are available for the optimal control of nonlinear stochastic systems. Although the optimal-control theory for deterministic systems has reached a respectable level of maturity, the corresponding theory for stochastic systems presents a serious challenge for practical implementation and meaningful applications.

An approximation theory, originally developed for deterministic control systems [Saridis and Lee (1979), Saridis and Balaram (1986), Wang and Saridis (1992)], may be used to find some feasible, practical solutions to the optimal control of nonlinear stochastic systems. The problem of stochastic control is addressed from an inverse point of view:

> Given an arbitrary selected admissible feedback control, find its measure with respect to a given performance cost, compare it to other feedback controls, and successively improve it to converge to the optimal.

This approach requires the extension of the Hamilton–Jacobi–Bellman theory to nonoptimal situations, where control variations are performed about a fixed but arbitrary control, as in the deterministic case.

5.2 PROBLEM FORMULATION

For the purpose of obtaining explicit expressions, but without loss of generality, because the results are immediately generalizable, consider a nonlinear stochastic control system described by the following stochastic differential equation:

$$dx = f(t,x)\,dt + b(t,x)u\,dt + g(t,x)\,dw, \qquad t \in I \equiv [t_0, T], \tag{5.1}$$

where $x \in R^n$ is a vector of state of the stochastic system, $u \in \Omega_u \subset R^m$ is a control vector, Ω_u is a specified compact set of admissible controls, and $w \in R^k$ is a separable Wiener process. $f\colon I \times R^n \to R^n$, $b\colon I \times R^n \to R^{n\times m}$, and $g\colon I \times R^n \to R^{n\times k}$ are measurable system functions.

For a given *initial* state $x(t_0) = x_0$ (deterministic) and feedback control $u(t, x)$, the performance cost of the system (5.1) is defined as

$$J(u; t_0, x_0) = E\left\{\int_{t_0}^{T}\left[L(t,x) + |u|^2\right] dt + \phi(T, x(T))/x(t_0) = x_0\right\}, \tag{5.2}$$

with nonnegative functions $L\colon I \times R^n \to R^1$ and $\phi\colon I \times R^n \to R^1$. J is also called the *value function* of system (5.1).

Equation (5.1) is written in a differential form, instead of the usual derivative one previously used, to emphasize the problems arising from the special characteristics of the Wiener process. One may formally divide by dt and replace dw/dt by $w(t)$, to recover the known form. This equation was studied first by Itô (1951), and later, under less restrictive conditions, by Doob (1953), Dynkin (1953), Skorokhod (1965), and Wang (1965). They have introduced a stochastic calculus that utilizes an *infinitesimal generator* $£_u$, which contains second-order derivative information, to replace the regular functional derivative, in order to account for the discrepancies of the derivatives of the Wiener process. This expression was encountered in the Hamilton–Jacobi–Bellman equation, when the process noise was modeled by a delta function. It is assumed that feedback control $u(t, x) \in \Omega_u$ satisfies the following conditions.

(i) Linear growth condition:

$$|f(t,x) + b(t,x)u(t,x)| + |g(t,x)| \le a(1 + |x|).$$

(ii) Uniform Lipschitz condition.

$$\|(f(t,x) - b(t,x)u) - (f(t,y) - b(t,y)u)\| + \|g(t,x) - g(t,y)\| \le a\|x - y\|,$$

where $\cdot(t, x)$, $\cdot(t, y) \in I \times R^n$, $\|\cdot\|$ is a Euclidian norm operator, and a is some constant.

The infinitesimal generator of the stochastic process specified by (5.1) is defined to be

$$\pounds_u\Psi \equiv \tfrac{1}{2}\,\mathrm{Tr}\big[g(t,x)g(t,x)^T\,\partial^2\Psi/\partial x^2\big] + \partial\Psi/\partial x^T\big[f(t,x) + b(t,x)u\big], \tag{5.3}$$

where $\Psi\colon I \times R^n \to R^1$ has compact support and is continuous up to all its second-order derivatives [Dynkin (1953)], and $(\cdot)^T$ and $\mathrm{Tr}(\cdot)$ are transpose and trace operators, respectively. $\partial\Psi/\partial t$ and $\partial\Psi/\partial x$ are the first-order time and space partial derivatives, respectively, and $\partial^2\Psi/\partial x^2$ is the second-order partial derivative of the function Ψ.

A pre-Hamiltonian function of the system with respect to the given performance cost (5.2) and a fixed but arbitrary control law $u(t, x)$ is defined as

$$H(x,\partial\Psi/\partial x,\partial^2\Psi/\partial x^2,u,t) = L(t,x) + \|u\|^2 + \pounds_u\Psi. \tag{5.4}$$

The optimal control of stochastic systems can be stated now.

Optimal Stochastic Control Problem. For a given initial condition $(t_0, x_0) \in I \times R^n$, and performance cost (5.2), find $u^* \in \Omega_u$ such that

$$V^*(t_0,x_0) \equiv J(u^*;t_0,x_0) = \inf_{u\in\Omega_u} J(u;t_0,x_0). \tag{5.5}$$

If it is assumed that the optimal-control law, $u(x,t)^*$ exists and if the corresponding value function $V^*(x,t)$ is sufficiently smooth, then u^* and V^* may be found by solving the well-known Hamilton–Jacobi–Bellman equation, defined for the optimal-control problem [Bellman (1956)],

$$V_t^* + \operatorname*{Min}_{u\in\Omega_u}\{\pounds_u V^* + L(t,x) + \|u\|^2\} = 0, \qquad V(T,x(T)) = \phi(T,x(T)). \tag{5.6}$$

Unfortunately, except in the case of linear quadratic Gaussian controls, where the problem has been solved [Wonham (1970)], a closed-loop form solution of the Hamilton–Jacobi–Bellman for solving the optimal stochastic control problem cannot be obtained in general when the system of Eqs. (5.1) is nonlinear.

Therefore, one may instead consider the optimal-control problem relaxed to that of finding an admissible feedback control law, $u(x, t)$, that has an acceptable (not necessarily optimal) value function. This gives rise to a *suboptimal stochastic control* that represents an *approximate solution*, and could conceivably be solved with less difficulty than the original stochastic optimal-control problem. A solution to this suboptimal stochastic control problem may be found by means of a successive approximation method to be discussed in the following.

5.3 APPROXIMATION THEORY OF STOCHASTIC OPTIMAL CONTROL

This section contains the main results of the approximation theory for the solution of nonlinear stochastic control problems. It will be established through the development of various theorems, which guarantee the rigor of the formulation. The proofs of these theorems are not included here. They can be found in Wang and Saridis (1992).

Two theorems, one for the evaluation of performance of control laws and the other for the construction of lower and upper bounds of value functions, are established first. Then theoretical procedures that can lead to the iterative design of suboptimal controls are developed based on those two theorems.

Theorem 5.1 (Performance Evaluation of Control Law) Assume V: $I \times R^n \to R^1$ be an arbitrary function with continuous V, $(\partial V/\partial t)$, $(\partial V/\partial x)$, and $(\partial^2 V/\partial x^2)$ that satisfy the condition

$$\|V\| + \|\partial V/\partial t\| + \|x\|\,\|\partial V/\partial x\| + \|x\|^2\|\partial^2 V/\partial x^2\| < b(1 + \|x\|^2), \quad (5.7)$$

where b is a suitable constant. Then the necessary and sufficient conditions for $V(t, x)$ to be the value function of an admissible fixed feedback control law $u(t, x) \in \Omega_u$, that is,

$$V(t, x) = E\left\{\int_t^T \left[L(t, x) + \|u\|^2\right] dt + \Phi(T, x(T))/x(t) = x\right\}, \qquad t \in I, \quad (5.8)$$

are

$$\partial V/\partial t + \pounds_u V + L(t, x) + \|u\|^2 = 0, \quad (5.9)$$

$$V(T, x(T)) = \Phi[T, x(T)]. \quad (5.10)$$

REMARK 1. For a feedback controller that makes system (5.1) unstable, the associated $V(t, x)$ does not satisfy the assumptions of Theorem 5.1, particularly, condition Eq. (5.7).

REMARK 2. The relation in (5.9) and (5.10), called the *generalized Hamilton–Jacobi–Bellman equation* for the stochastic control system (5.1), is reduced to the Hamilton–Jacobi–Bellman partial differential equation (5.6) when the control $u(x, t)$ is replaced by the optimal control $u^*(x, t)$.

REMARK 3. The exact value function for a given control law $u(x, t)$ is found by solving the *generalized Hamilton–Jacobi–Bellman* partial differential equation. This equation, though simpler and more general than the Hamilton–Jacobi–Bellman equation, is difficult to solve in a closed form.

REMARK 4. For more general stochastic processes defined by the stochastic differential equation $dx = f(t, x, u)\,dt + g(t, x)\,dw$ and the general form of performance cost, one can show that Theorem 1 is still true.

Because it is generally difficult to find the exact value functions satisfying Eqs. (5.9) and (5.10) of Theorem 5.1, the following theorem introduces a method of constructing the lower and upper bounds of value functions. This method can be used to design simpler suboptimal controllers based only on the upper bounds of the value functions.

Theorem 5.2 (Lower and Upper Bounds of Value Function) Assume an admissible fixed feedback control law $u(t, x) \in \Omega_u$, and a continuous function $s(t, x)$, with $|s(t, x)| < \infty$ for all $(t, x) \in I \times R^n$. If the function $V(t, x)$ satisfies Eq. (5.7) with continuous V, $\partial V/\partial t$, $\partial V/\partial x$, $\partial^2 V/\partial x^2$, and

$$\partial V/\partial t + £_u V + L(t, x) + \|u\|^2 \equiv \nabla V \leq s(t, x) \leq 0 \qquad [\geq s(t, x) \geq 0], \tag{5.11}$$

$$V(T, x(T)) \geq \Phi(T, x(T)) \qquad (\leq \Phi(T, x(T))), \tag{5.12}$$

then $V(t, x)$ is an upper (or a lower) bound to the value function of system (5.1). That is,

$$V(t, x) \geq J(u; t, x) \qquad (\leq J(u; t, x)), \forall\, (t, x) \in I \times R^n. \tag{5.13}$$

REMARK 1. In general, the function $s(t, x)$ in this theorem does not need to be calculated. However, stating the inequality as in (5.11) gives an additional degree of flexibility that enables the determination of an upper (or a lower) bound to the value function $J(u;\, t, x)$.

REMARK 2. The function $V(t, x)$ in this theorem is the exact value function for a system with an augmented performance index by $s(t, x)$,

$$I(u; t_0, x_0) = E\left\{\int_{t_0}^{T}\left[L(t, x) + \|u\|^2 - s(t, x)\right] dt - \Phi_s(T, x(T))/x(t_0) = x_0\right\}, \tag{5.14}$$

where $\Phi_s(\cdot, \cdot)$ is a terminal cost function such that $\Phi_s(T, x(T)) \geq (\leq)\Phi(T, x(T))$.

Having established the two theorems for the evaluation of performance of a given feedback control law, it is necessary to develop algorithms to improve the control law. The following Theorems 5.3–5.5 provide a theoretical procedure for designing the suboptimal feedback controllers based on Theorem 5.1, and Theorem 5.6 presents a method for constructing upper and lower bounds to the optimal value function, which can be used to evaluate the acceptability of suboptimal controllers.

Theorem 5.3 Given the admissible controls $u_1 \in \Omega_u$ and $u_2 \in \Omega_u$, with $V_1(t, x)$ and $V_2(t, x)$, the corresponding value functions satisfying Eqs. (5.7) and (5.8) for u_1 and u_2, respectively, define the Hamiltonian functions for $i = 1, 2$:

$$H_{i\,\min} = H\left(x, \partial V_i/\partial x, \partial^2 V_i/\partial x^2, u_i^\circ, t\right) = L(t, x) + \|u_i^\circ\|^2 + \pounds_{u_i^\circ} V_i, \tag{5.15}$$

where

$$u_i^\circ(t, x) = -\tfrac{1}{2} b(t, x)^T \, \partial V_i(t, x)/\partial x. \tag{5.16}$$

It is shown that

$$V_i \geq V_2 \tag{5.17}$$

when

$$\partial V_1/\partial t + H_{1\,\min} \leq \partial V_2/\partial t + H_{2\,\min}. \tag{5.18}$$

REMARK. One should not try to find $\Delta V(t, x)$ by subtracting $V_1(t, x)$ from $V_2(t, x)$ directly based on their individual Itô formulas evaluated along the trajectories generated by their corresponding controls. This is due to the fact that the two state trajectories generated by $u_1(t, x)$ and $u_2(t, x)$ are different.

A combination of Theorems 5.1 and 5.3, in which case a $V(t, x)$ represents the value function of the system (5,1) when driven by control $u(t, x)$, yields an inequality that, as a basis of suboptimal control algorithms, iteratively reduces the cost of the performance of the system. This is outlined by the following theorem.

Theorem 5.4 Assume that there exists a control $u_1 \in \Omega_u$ and a corresponding function $V_1(t, x)$ satisfying Eqs. (5.7) and (5.8) of Theorem 1. If there exists a function $V_2(t, x)$ satisfying the same conditions of Theorem 5.1, and its associated control $u_2 \in \Omega_u$, has been selected to satisfy

$$\left\| u_2 + \tfrac{1}{2} b^T \partial V_2/\partial x \right\| \leq \left\| u_1 + \tfrac{1}{2} b^T \partial V_1/\partial x \right\|, \tag{5.19}$$

then,

$$V_1 \geq V_2. \tag{5.20}$$

REMARK. Clearly, condition (5.19) in Theorem 5.4 is much easier to test than condition (5.18) in Theorem 5.3, because (5.19) does not involve the time derivative $\partial V/\partial t$, and the infinitesimal generator $£_u V$.

Based on Theorems 5.3 and 5.4, the following theorem establishes a sequence of feedback controls that successively improve the performance cost, and converge to the optimal feedback control.

Theorem 5.5 Let a sequence of pairs $\{u_i, V_i\}$ satisfy (5.7)–(5.8) of Theorem 5.1, and u_i be obtained by minimizing the pre-Hamiltonian function corresponding to the previous value function V_{i-1}, that is,

$$u_i = -\tfrac{1}{2} b^T \partial V_{i-1}/\partial x, \qquad i = 1, 2, \ldots. \tag{5.21}$$

Then the corresponding value functions V_i satisfy the inequality

$$V_{i-1} \geq V_i, \qquad i = 1, 2 \ldots.$$

Thus, by selecting the pairs $\{u_i, V_i\}$ in the above manner sequentially, the resulting sequence $\{V_i\}$ converges monotonically to the optimal value function V^*, and the corresponding sequence $\{u_i\}$ converges to the optimal control u^*, associated with V^*.

REMARK 1. It follows from this theorem that the optimal feedback control u^* and the optimal value function V^* are related by

$$u^*(t, x) = -\tfrac{1}{2} b^T(t, x)\, \partial V^*(t, x)/\partial x, \qquad \forall\, (t, x) \in I \times R^n,$$

which is a relationship that results from the minimization of the Hamiltonian function associated with the stochastic system (5.1).

REMARK 2. As indicated by the conditions, in applying the theorem, assumptions must be made a priori regarding the admissibility of the successively derived control laws and their corresponding value functions. However, for a nonlinear stochastic control system as in (5.1), the admissibility of the new control laws is very difficult to show in general.

Finally, the following theorem presents a method for the construction of an upper (or a lower) bound of the optimal value function $V^*(t, x)$. Because the optimal value function itself is extremely difficult to find, its upper (or lower) bounds can provide a practical measure to evaluate the effectiveness of the suboptimal controllers.

Theorem 5.6 Assume that there exists a function $V^s(t, x)$ satisfying condition (5.7) of Theorem 5.1, for which the associated control

$$u^s(x,t) = -\tfrac{1}{2}b^T(t,x)\,\partial V^s(t,x)/\partial x \tag{5.22}$$

is an admissible one. Then $V^s(t, x)$ is an upper (or a lower) bound to the optimal value function $V^*(t, x)$ of system (5.1), if it satisfies the following conditions:

$$\partial V^s/\partial t + £_u V^s + L(t,x) + \|u^s\|^2 = s(t,x) \leq 0 \qquad (\geq 0), \tag{5.23}$$

$$V^s(T, x(T)) \geq \phi(T, x(T)) \qquad (\leq \Phi(T, x(T))), \tag{5.24}$$

where $s(t, x)$ is continuous and $|s(t, x)| < \infty$, for all $(t, x) \in I \times R^n$.

Theorems, which can lead to the design of simpler suboptimal controllers based on the upper and lower bounds of the value functions, can also be constructed. More detailed discussion regarding this for the stochastic regulator problem with infinite-time is given in the next section.

5.4 STOCHASTIC INFINITE-TIME REGULATOR

The stochastic regulator problem with infinite-time is defined as a control problem for the nonlinear stochastic system (5.1) with infinite duration $T \to \infty$. All state trajectories generated by admissible controls in Ω_u must be bounded uniformly in $I \times R^n$.

For the stochastic regulator problem with infinite horizon, the performance index of the system is defined to be

$$J(u; t_0, x_0) = \lim_{T\to\infty} E\left\{(1/T)\int_{t_0}^{T}\left[L(t,x) + \|u\|^2\right] dt/x(t_0) = x_0\right\}, \tag{5.25}$$

and Itô's integration formula applied to this case becomes

$$V(t,x) = -E\left\{\int_t^T [\pounds_u V(\tau, x(\tau)) + \partial V/\partial\tau(\tau, x(\tau))]\, d\tau / x(t) = x\right\}, \qquad t \in I,$$

where $V(t,x)$ satisfies $V(t,0) = 0$ and (5.8) of Theorem 5.1 for all the possible state trajectories, which is true for all $u \in \Omega_u$ [Wang (1965)]. All the theorems developed in the previous section are still valid for the stochastic regulator problem with infinite horizon, except that all the terminal conditions at $t = T$ in those theorems are no longer required. However, in this case, theorems can be constructed that can lead to the iterative design of simpler suboptimal controls based only on the upper and lower bounds to value functions. Because in general the upper and lower bounds can be obtained without solving the partial differential equation (5.9), those theorems have a great potential for application. Three such theorems, corresponding to Theorems 5.3–5.5, follow.

Theorem 5.7 Given admissible controls u_1 and $u_2 \in \Omega_u$, with $J_1(t,x)$ and $J_2(t,x)$ their corresponding value functions defined by (5.25), if there exist function pairs $V_1(t,x)$, $s_1(t,x) \geq 0$, and $V_2(t,x)$, $s_2(t,x) \leq 0$ satisfying (5.11) of Theorem 5.2 for u_1 and u_2, respectively, then

$$J_1 \geq J_2 \tag{5.26}$$

when

$$\partial V_1/\partial t + H_{1\,\min} \leq \partial V_2/\partial t + H_{2\,\min}. \tag{5.27}$$

The next theorem is the counterpart of Theorem 5.4.

Theorem 5.8 Assume that there exist a control $u_1 \in \Omega_u$ and a function pair $\{V_1(t,x), s_1(t,x) \geq 0\}$ (5.11) of Theorem 5.2. If there exists a function pair $\{V_2(t,x), s_2(t,x) \leq 0\}$ satisfying the same condition of Theorem 5.2, of which the associated control $u_2 \in \Omega_u$, has been selected to satisfy

$$\left\| u_2 + \tfrac{1}{2} b^T \partial V_2/\partial x \right\|^2 \leq \left\| u_1 + \tfrac{1}{2} b^T \partial V_1/\partial x \right\|^2, \tag{5.28}$$

then

$$J_1 \geq J_2 \tag{5.29}$$

where $J_1(t,x)$ and $J_2(t,x)$ are the value functions of u_1 and u_2, respectively.

Theorem 5.9 This theorem is a direct extension of Theorem 5.5, based on Theorem 5.8. The pairs $\{u_i, V_i\}$ satisfy (5.7) and (5.8) of Theorem 5.1, and u_i minimizes the pre-Hamiltonian corresponding to the previous value function V_{i-1}.

5.5 STABILITY CONSIDERATIONS

Stability of the inifinite-time approximate control problem will be treated in a manner similar to that of Section 4.8, *it suffices to show that the performance index (5.25) of the system (5.1) is bounded for all the controls generated by the approximation theory.*

Lemma 5.1. If Theorems 5.7 and/or 5.8 are satisfied, stability of the infinite-horizon systems, driven by the subsequent controls generated by the approximation theory, is generated if the first controller u_1 is selected to yield a bounded J_1.

The proof of the above lemma is clearly established from the statements of Theorems 5.7 and/or 5.8. In order to prove the stability of the system for the first controller the following steps are appropriate for consideration:

1. System 5.1 must be completely controllable (CC), for all admissible controls u.
2. Because all the states are assumed available for measurement, system (5.1) is obviously completely observable (CO).
3. The performance index (5.25) is bounded because

$$\begin{aligned} J(u;t,x) &= \lim_{T\to\infty} E\left\{(1/T)\int_t^T\left[L(\tau,x)+\|u\|^2\right]d\tau/x(t)=x\right\} \\ &= \lim_{T\to\infty}(1/T)\left[V(t,x)\right]. \end{aligned}$$

From Itô's integration formula,

$$V(t,x) = -E\left\{\int_t^T\left[£_u V(\tau,x(\tau))+\partial V/\partial\tau\right]d\tau/x(t)=x\right\},$$

where

$$\begin{aligned} £_u V(t,x(t)) &= \tfrac{1}{2}\operatorname{Tr}\left[g(t,x)g(t,x)^T\,\partial^2 V/\partial x^2\right] \\ &\quad + \partial V^T/\partial x\left[f(t,x)+b(t,x)u\right]. \end{aligned}$$

Select the first control u_1 to satisfy the previous theorems of the approximation theory, and the condition

$$E\left[£_u V(\tau,x(\tau))+V_\tau\right]\le\infty, \qquad \forall\,\tau\in[t,\infty].$$

Then using the mean value theorem,

$$V(t,x)=E\left[£_u V(\alpha,x(\alpha))+V_\alpha\right](T-t)\le M(T-t), \qquad \alpha\in[t,T].$$

Then

$$
\begin{aligned}
J(u; t, x) &= \lim_{T\to\infty} E\left\{(1/T)\int_t^T \left[L(\tau, x) + \|u\|^2\right] d\tau / x(t) = x\right\} \\
&= \lim_{T\to\infty} 1/T[V(t, x)] \\
&= \lim_{T\to\infty} [M(T - t)/T] \le M < \infty. \qquad (5.30)
\end{aligned}
$$

The boundedness of the performance index $J(u;\, t, x)$ of the infinite-horizon problem establishes the stability of the system for all the controllers derived from the approximation theory.

5.6 DESIGN OF SUBOPTIMAL CONTROLLERS

The optimal feedback control $u^*(t, x)$ and its associated $V^*(t, x)$ satisfying the Hamilton–Jacobi–Bellman equation, Eq. (5.6), obviously satisfy all the theorems developed in Section 5.3. However, in most cases of nonlinear stochastic control systems, the optimal solution is very difficult, if not impossible, to implement, either because the solution is unavailable or because some of the states are not available for measure. In both cases the theory developed in Section 5.3 may serve to obtain controllers that can make the system stable, and then be successively modified to approximate the optimal solution.

Upper and lower bounds of the value function of the nonlinear stochastic system may be used to evaluate the effectiveness of the approximation, and provide a yardstick for its success.

5.6.1 Exact Design Procedure

This approach, based on the assumption that the value function $V(t, x)$ for a control $u(t, x)$ can be found to satisfy (5.9) and (5.10) of Theorem 1, may be implemented according to Theorem 5.3 and 5.4, by the following procedure to design a control, for system (5.1):

1. Select a feedback control law $u_0(t, x)$ for system (5.1); set $i = 0$.
2. Find a $V_i(t, x)$ to satisfy Theorem 5.1 for $u_i(t, x)$.
3. Obtain a $u_{i+1}(t, x)$ and a $V_{i+1}(t, x)$ to satisfy Theorem 5.1 and Theorems 5.3 or 5.4 for u_i and V_i: u_{i+1} is an improved controller.
4. From Theorem 5.6, find a lower bound $V_L(t, x)$ to the optimal value function $V^*(t, x)$, and then use $V_{i+1} - V_L$ as a measure to evaluate $u_{i+1}(t, x)$ as an approximation to the optimal control $u^*(t, x)$. If acceptable, stop.
5. If the approximation is not acceptable, repeat Step 2 by increasing index i by 1 and continue.

The improved controller u_{i+1} in Step 3 can also be constructed by using Theorem 5.5, if the corresponding value function V_{i+1} can be obtained. When a lower bound to the optimal value function is difficult to find, step 4 can be omitted and then the criterion for the acceptability of the approximation has to be determined based on other considerations.

EXAMPLE 1 (Linear Stochastic Systems) As the first example, the design procedure is applied to a linear stochastic system described by

$$dx = A(t)x\,dt + B(t)u\,dt + G(t)\,dw.$$

The cost function of the system has the quadratic form

$$J(u; t_0, x_0) = E\left\{\int_{t_0}^{T}\left[x^T M(t)x + \|u\|^2\right]dt + x(T)^T \Phi x(T)/x(t_0) = x_0\right\}.$$

The infinitesimal generator of the linear stochastic process is

$$£_u V(t, x) = \tfrac{1}{2}\operatorname{Tr}\left[G(t)G(t)^T \partial^2 V/\partial x^2\right] + \partial V^T/\partial x\left[A(t)x + B(t)u\right].$$

Assume first a linear control,

$$u_1(t, x) = -K_1(t)x,$$

where $K_1(t)$ is a feedback matrix. The corresponding value function is assumed to be

$$V_1(t, x) = s_1(t) + x^T S_1(t)x,$$

where s_1 and S_1 can be found by solving (5.9) and (5.10) of Theorem 5.1,

$$dS_1/dt + (A - BK_1)^T S_1 + S_1^T(A - BK_1) + M + K_1^T K_1 = 0, \qquad S_1 = \Phi,$$

and

$$s_1(t) = \int_t^T \operatorname{Tr}\left[G(\tau)G(\tau)^T S_1(\tau)\right]d\tau.$$

The feedback law is improved by using Theorem 5.5. From Eq. (5.24),

$$u_i(t, x) = -\tfrac{1}{2}B^T(t)\partial V_{i-1}/\partial x = -B^T S_{i-1}(t)x = -K_i(t)x, \qquad i \geq 2,$$

and the corresponding value function is assumed to be

$$V_i(t, x) = s_i(t) + x^T S_i(t)x, \qquad i \geq 2,$$

where s_i and S_i are determined by solving the equations

$$dS_i/dt + (A - BK_i)^T S_i + S_i^T(A - BK_i) + M + K_i^T K_i = 0. \qquad S_i(T) = \Phi,$$

$$s_i(t) = \int_t^T \mathrm{Tr}\left[G(\tau)G(\tau)^T S_i(\tau)\right] d\tau.$$

As $i \to \infty$ and $S_i(t)$ approaches S, the solution of the matrix Riccati equation is,

$$dS/dt + (A - BK)^T S + S^T(A - BK) + M + K^T K = 0,$$

$$S(T) = \Phi, \qquad K = B^T S,$$

and correspondingly, the control approaches

$$u(t, x) = -B(t)^T S(t)x = -K(t)x,$$

which is the optimal control for the linear stochastic systems with quadratic performance criterion [Wonham (1970)].

EXAMPLE 2 The second example illustrates the design method for the following nonlinear first-order stochastic system:

$$dx = -x\,dt + u\,dt + \tfrac{1}{2}x\,dw, \qquad x(0) = x_0,$$

with a cost function,

$$J(u; t_0, x_0) = E\left\{\int_{t_0}^{\infty} [10x^2 + x^4 + u^2]\,dt/x(t_0) = x_0\right\}.$$

The infinitesimal generator of the stochastic process becomes

$$£_u V(t, x) = \tfrac{1}{4}x^2 \partial^2 V/\partial x^2 + (-x + u)\partial V/\partial x,$$

First assume a *linear control law*,

$$u_1(x) = -ax, \qquad a > 0.$$

The corresponding value function is assumed to be

$$V_1(x) = s_1 x^2 + s_2 x^4.$$

Equation (5.9) of Theorem 5.1 yields

$$\left[10 + a^2 - s_1\left(2a + \tfrac{7}{4}\right)\right]x^2 + \left[1 - s_2\left(4a + \tfrac{5}{2}\right)\right]x^4 = 0,$$

which is true for

$$s_1 = (40 + 4a^2)/(8a + 7), \qquad s_2 = 2/(8a + 5), \qquad a > 0.$$

Next, select a higher-order control law,

$$u_2(x) = -ax - bx^3, \qquad b > 0.$$

The corresponding value function is assumed to be

$$V_2(x) = q_1 x^2 + q_2 x^4.$$

In this case, in order to satisfy (5.9), one must solve

$$\begin{aligned}&\left[10 + a^2 - q_1\left(2a + \tfrac{7}{4}\right)\right]x^2 \\ &\qquad + \left[1 + 2ab - 2bq_1 - q_2\left(4a + \tfrac{5}{2}\right)\right]x^4 + \left(b^2 - 4bq_2\right)x^6 = 0,\end{aligned}$$

which is true for

$$\begin{aligned}q_1 = (40 + 4a^2)/(8a + 7), \qquad q_2 = (16a + 14)/(675 - 16a), \qquad b = 4q_2, \\ 0 < a < 675/16 = 42.1875.\end{aligned}$$

To satisfy Theorem 5.4, controllers u_1 and u_2 must satisfy

$$\|u^2 + q_1 x + 2q_2 x^3\| \leq \|u_1 + s_1 x + 2s_2 x^3\|,$$

which yields

$$0 < a \leq (\sqrt{689}\} - 7)/8 = 2.4061.$$

Their corresponding value functions can be compared:

$$\begin{aligned}V_1(x) &= (40 + 4a^2)/(8a + 7)x^2 + 2/(8a + 5)x^4, \\ V_2(x) &= (40 + 4a^2)/(8a + 7)x^2 + (16a + 14)/(675 - 16a)x^4, \\ \Delta V(x) &= V_2(x) - V_1(x) \\ &= 32(4a^2 + 7a - 40)/(8a + 5)(675 - 16a)x^4 \geq 0.\end{aligned}$$

Figure 5.1 illustrates V_1, V_2, and the corresponding performance improvement for $a = 0.2723$.

If Theorem 5.5 is to be used for the above u_1 and V_1, u_2 must be selected according to

$$u_2(x) = -\tfrac{1}{2}\partial V_1/\partial x = -s_1 x - 2s_2 x^3,$$

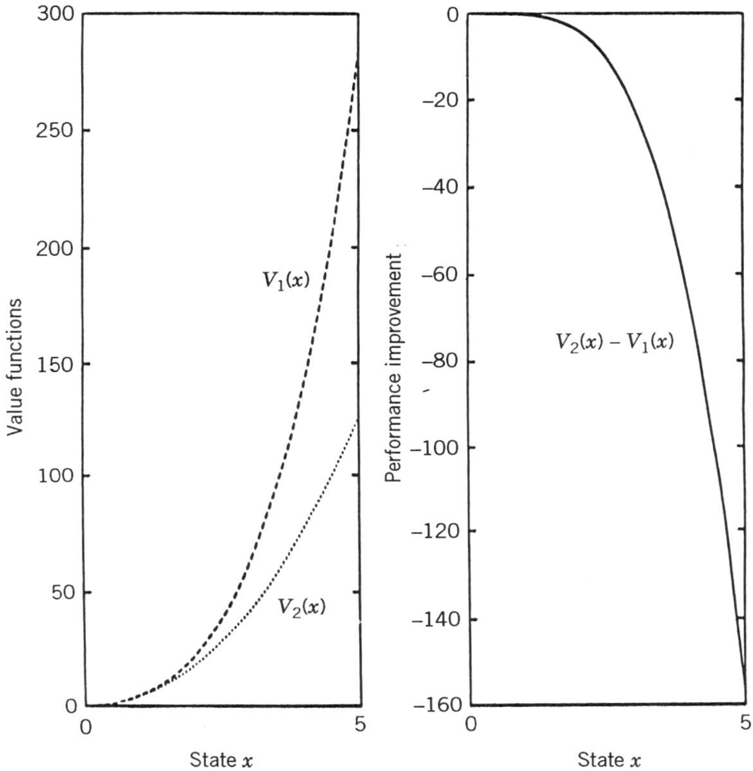

Fig. 5.1 Example 2; design with Theorem 5.4

and a V_2 satisfying (5.9) of Theorem 5.1 exists if

$$\begin{aligned} V_2(x) &= q_1 x^2 + q_2 x^4, \\ q_1 &= (40 + 4s_1^2)/(8s_1 + 7) = 2.7800, \\ q_2 &= (16s_1 + 14)/(675 - 16s_1) = 0.1393, \\ a &= 0.2723, \qquad s_1 = 4.3904. \end{aligned}$$

Comparing the value functions, one finds that

$$\begin{aligned} V_1(x) &= 4.3904x^2 + 0.2786x^4, \\ V_2(x) &= 2.7800x^2 + 0.1393x^4, \\ \Delta V(x) &= V_2(x) - V_1(x) = -1.6104x^2 - 0.1393x^4. \end{aligned}$$

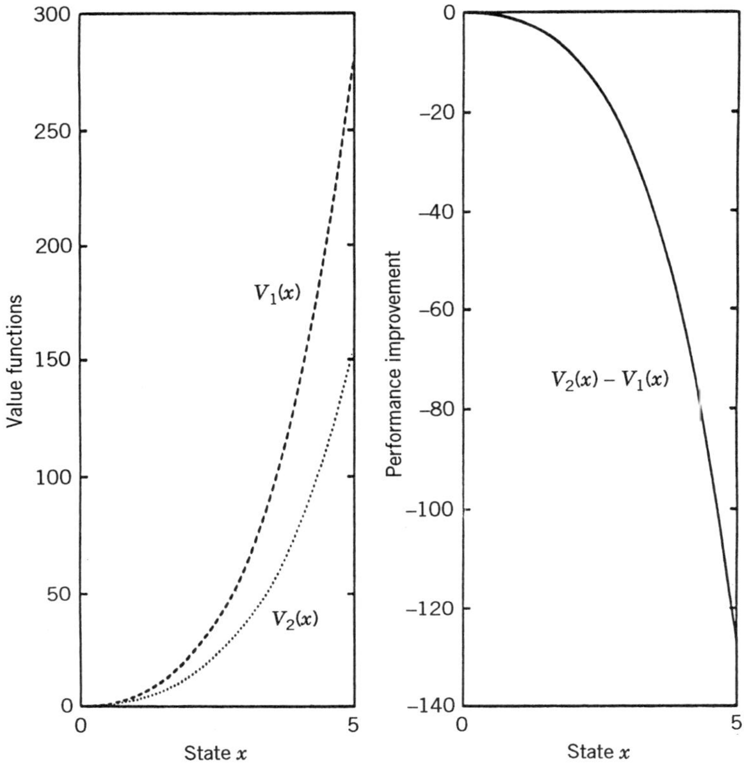

Fig. 5.2 Example 2; design with Theorem 5.5

V_1, V_2, and the performance improvement are shown in Figure 5.2. Comparing Figure 5.2 with Figure 5.1, one can see that the performance improvement in this case is not as big as that made by the design using Theorem 5.4.

In both cases, because $\Delta V \leq 0$, the performance of the system has been improved by replacing a linear controller u_1 by a nonlinear controller u_2. All the above controllers make the origin an equilibrium point for the system.

5.6.2 Approximate Design for Regulator Problems

In many cases the selection of a $V(t, x)$ to satisfy (5.9) and (5.10) of Theorem 5.1 is a very difficult task. In such a case approximate design procedures, which use the upper and lower bounds of the value function obtained through Theorem 5.2, can be constructed. For the infinite-time stochastic regulator problem, the following design procedure is proposed based on

Theorem 5.7 and 5.8 in Section 5.4:

1. Select a feedback control law $u_0(t, x)$ for system (5.1); set $i = 0$.
2. For a $s_i(t, x) \geq 0$, find a $V_i(t, x)$ for u_i to satisfy Theorem 5.2 for a lower bound.
3. Obtain a $u_{i+1}(t, x)$, and for an $s_{i+1} \leq 0$, and find a $V_{i+1}(t, x)$ for u_{i+1} to satisfy Theorem 2 as an upper bound. $u_{i+1}(t, x)$, s_{i+1}, and $V_{i+1}(t, x)$ found, should also satisfy conditions (5.40) of Theorem 5.7 or (5.41) of Theorem 5.8 for the improvement of performance.
4. Using a lower bound to the optimal value function, which is determined according to Theorem 5.6, the approximation of the optimal control can be measured, If acceptable, stop.
5. If the approximation is not acceptable, repeat Step 2 by increasing index i by 1 and continue.

EXAMPLE 3 The design method is illustrated with the following nonlinear first-order stochastic regulator problem,

$$dx = x^3\,dt + u\,dt + \tfrac{1}{2}x\,dw, \qquad x(t_0) = x_0,$$

with a value function,

$$J(u; t_0, x_0) = E\left\{\int_{t_0}^{\infty}[10x^2 + x^4 + u^2]\,dt/x(t_0) = x_0\right\}.$$

The infinitesimal generator of the stochastic process is

$$£_u V(t, x) = \tfrac{1}{4}x^2\partial^2 V/\partial x^2 + (x^3 + u)\partial V/\partial x,$$

For a *linear control law*,

$$u_1(x) = -a_1 x, \qquad a_1 > 0.$$

The lower bound of its value function is assumed to be

$$V_1(x) = s_1 x^2 + s_2 x^4.$$

Application of (5.11) of Theorem 5.2 leads to

$$\left[10 + a_1^2 - s_1\left(2a_1 - \tfrac{1}{4}\right)\right]x^2 + \left[1 + 2s_1 - s_2\left(4a_1 - \tfrac{3}{2}\right)\right]x^4 + 4s_2x^6 \geq 0,$$

which is satisfied by

$$s_1 = \left[40(1 - c_1) + 4a_1^2\right]/(8a_1 - 1),$$
$$s_2 = \left[2(1 - c_2) + 4s_1\right]/(8a_1 - 3), a_1 > \tfrac{3}{8}, \quad \text{for any } 1 \geq c_1, \quad c_2 > 0.$$

For a *higher-order control law*,

$$u_2(x) = -a_2 x - b_2 x^3, \qquad a_2 > 0, \qquad b_2 > 0,$$

the upper bound to its value function is assumed to be

$$V_2(x) = q_1 x^2 + q_2 x^4.$$

In this case, application of (5.11) yields

$$\left[10 + a_2^2 - q_1(2a_2 - \tfrac{1}{4})\right]x^2 + \left[1 + 2q_1 - 2(q_1 - a_2)b_2 - q_2(4a_2 - \tfrac{3}{2})\right]x^4$$
$$+ \left[b_2^2 - 4q_2(b_2 - 1)\right]x^6 \leq 0,$$

which is true for

$$q_1 = \left[40(1 + d_1) + 4a_2^2\right]/(8a_2 - 1), \qquad q_2 = (1 + d_2)b_2^2/(4b_2 - 4),$$
$$1 + 2q_1 - 2(q_1 - a_2)b_2 - q_2(4a_2 - \tfrac{3}{2}) \leq 0, \qquad a_2 > \tfrac{3}{8}, \qquad b_2 > 1,$$

where $d_1, d_2 > 0$ are arbitrary.

Improvement of performance $\Delta V \leq 0$ occurs if (5.28) or (5.29) is satisfied, which leads to,

$$s_1 = \left[40(1 - c_1) + 4a_1^2\right]/(8a_1 - 1) \geq q_1 = \left[40(1 + d_1) + 4a_2^2\right]/(8a_2 - 1),$$
$$s_2 = \left[2(1 - c_2) + 4s_1\right]/(8a_1 - 3) \geq q_2 = (1 + d_2)b_2^2/(4b_2 - 4),$$

which, with the rest of the inequalities, produce acceptable values for a_1, a_2, and b_2. For example, one can show that

$$a_1 = 10, \qquad a_2 = \tfrac{3}{2}, \qquad b_2 = 2, \qquad c_1 = c_2 = d_1 = d_2 = 0.1$$

is a set of acceptable values. The lower and upper bounds of the value functions in this case are found as

$$V_1(x) = 5.5190x^2 + 0.3101x^4,$$
$$V_2(x) = 4.8182x^2 + 0.0688x^4,$$
$$\Delta V(x) = V_2(x) - V_1(x) = -0.7008x^2 - 0.2413x^4.$$

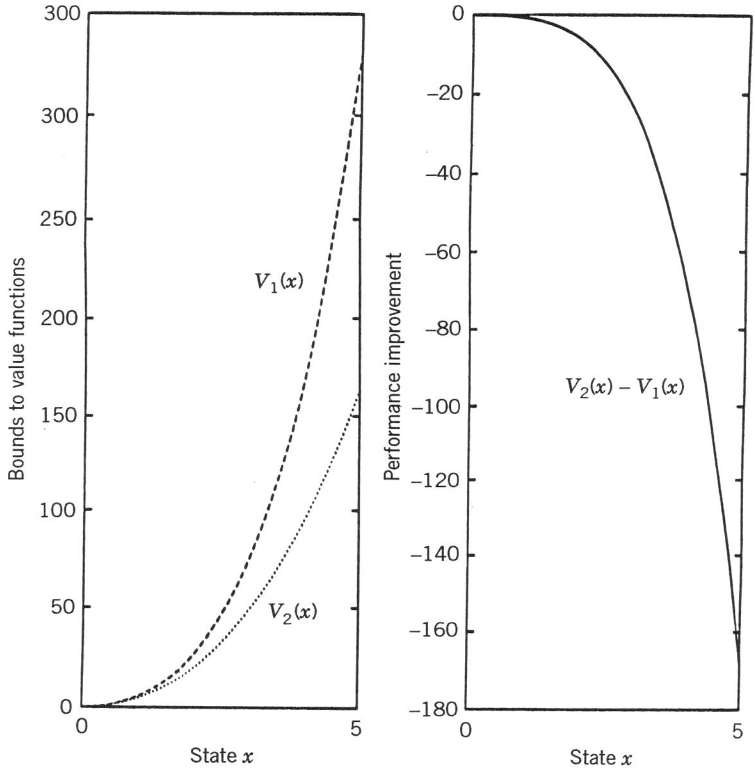

Fig. 5.3 Example 3; design with Theorem 5.8

The results are illustrated in Figure 5.3. Again, improvement of performance occurs by replacing a linear controller u_1 by a nonlinear controller u_2. Note that in this case the actual value function of control u_2 cannot be found by using the method applied in Example 2. This approach has a great potential for application, because one does not have to solve the partial differential equation of Theorem 5.1 every time.

5.7 CONCLUSIONS

An approximation theory of optimal control for nonlinear stochastic systems has been developed. The theory demonstrates the following points.

1. The iteration scheme and the theorem for the construction of upper and lower bounds to value functions, proposed by Saridis and Lee (1979) for deterministic systems, is generalized to the case of stochastic systems. Extensions to the discrete-time case can be easily obtained.

2. A successive design procedure using upper and lower bounds to the exact value functions has been developed for the stochastic regulator problem with infinite horizon. The determination of the upper and lower bounds requires the solution of a partial differential inequality instead of an equality. Therefore it provides a degree of flexibility in the design method over the exact design method.

Several examples are used to illustrate the application of the proposed approximation theory to stochastic control. It has been shown that in the case of linear quadratic Gaussian problems, the approximation theory leads to the exact solution of optimal control. The major drawback of the established theory lies in that the stability of suboptimal controllers has been assumed a priori, and not guaranteed by, the design procedure. This is not surprising, because the stability problem is very difficult for stochastic control systems, and very few results are available so far. Further study is required to generate stable sequences for the design of successively improving feedback controllers. Moreover, for the actual implementation of the design procedure, the computational problem has to be addressed.

PROBLEMS

5.1 Given the nonlinear stochastic system, with fixed initial condition $x(0) = x_0$, and process noise $w(t) \sim N(0, q^2)$,

$$dx/dt = -x(t) + u(t) + x(t)w(t),$$

and the cost function

$$J(u) = E\left\{ \int_0^\infty (4x^2 + 2x^4 + 2u^2)\, d\tau / x(0) = x_0 \right\},$$

(a) find the cost $V_1(x) = J(u_1)$ for $u_1(x) = -x$, and

(b) find the controller $u_2(x)$ that improves the cost function $V_2(x) = J(u_2)$ satisfying Theorem 5.5 of the approximation theory, without solving for $V_2(x)$.

REFERENCES

Aoki, M., *Optimization of Stochastic Systems*, Academic Press, New York, 1967.

Bellman, R., *Dynamic Programming*, Princeton University Press, Princeton, NJ, 1956.

Doob, J. L., *Stochastic Processes*, Wiley, New York, 1953.

Dynkin, E. B., *Markov Processes*, Academic Press, New York, 1953.

Itô, K., *On Stochastic Differential Equations*, *Mem. Am. Math. Soc.*, Vol. 4, 1951.

Kwakernaak, H., and Sivan, R., *Linear Optimal Control Systems*, Wiley, New York, 1972.

Sage, A. P., and White, C. C., *Optimum Systems Control*, Prentice-Hall, Englewood Cliffs, NJ, 1977.

Saridis, G. N., and Lee, C. S. G., "An Approximation Theory of Optimal Control for Trainable Manipulators," *IEEE Trans. Syst., Man, Cybern.*, Vol. SMC-9, No. 3, pp. 152–159, 1979.

Saridis, G. N., and Balaram, J., "Suboptimal Control for Nonlinear Systems," *Control-Theory and Advanced Technology*, MITA Press, Tokyo, Japan, 1986, Vol. 2, No. 3, pp. 547–562.

Skorokhod, A. V., *Studies in the Theory of Random Processes*, Addison-Wesley, Reading, MA, 1965.

Wang, Z. K., *Theory of Stochastic Processes*, Academic Press, Beijing, 1965.

Wang, F. Y., and Saridis, G. N., "Suboptimal Control for Nonlinear Stochastic Systems," *Proceedings of the 1992 Conference on Decision and Control*, Tucson, AZ, Dec. 1992.

Wonham, W. M., "Random Differential Equations in Control Theory," in A. T. Bharucha-Reid (Ed.), Probabilistic Methods in Applied Mathematics, Academic Press, New York, 1970.

6

ENTROPY REFORMULATION OF STOCHASTIC ESTIMATION AND OPTIMAL CONTROL

6.1 REVIEW OF ENTROPY CONCEPTS

Entropy is a form of low-quality energy, first encountered in thermodynamics. It represents an undesirable form of energy that is accumulated when any type of work is generated. Recently it served as a model of different types of energy-based resources, like transmission of information, biological growth, environmental waste, et cetera. Entropy was recently introduced as a unifying measure of performance of the different levels of an Intelligent Machine by Saridis (1985b). Such a machine is aimed at the creation of *modern intelligent robots that may perform human tasks with minimum interaction with a human operator*. Because the activities of such a machine are energy related, entropy may easily serve as a cost measure of performing various tasks, such as Intelligent Control, Image Processing, Task Planning and Organization, and System Communication among diversified disciplines with different performance criteria. The model to be used is borrowed from Information Theory, where the uncertainty of design is measured by a probability density function over the appropriate space, generated by Jaynes' Maximum Entropy Principle. Other applications of the entropy concept are defining reliability measures for design purposes and obtaining measures of complexity of the performance of a system, useful in the development of the theory of intelligent machines.

Entropy is a convenient global measure of performance because of its wide applicability to a large variety of systems of diverse disciplines, including waste processing, environmental, socioeconomic, biological, and other. Thus, by serving as a common measure, it may expand system integration by incorporating, say, societal, economic, or even environmental systems to engineering processes.

In this chapter, entropy concepts will be mainly used for the reformulation of the theory of stochastic optimal control and its approximation theory, state estimation, and parameter identification used in adaptive control systems. It will prove Fel'dbaum's claim that the stochastic Open-loop-feedback-optimal and adaptive controls, based on the *certainty equivalence principle* [Kumar and Varaiya (1986), Morse (1990), (1992)] are not optimal. It will also provide a measure of goodness of their approximation to the optimal in the form of an upper bound of the entropy missed by the approximation.

An application to the design of Intelligent Machines using performance measures based on reliability and complexity will demonstrate the power of the approach.

6.1.1 Entropy and Thermodynamics

The concept of entropy was introduced in thermodynamics by Clausius in 1867, as the low-quality energy resulting from the second law of thermodynamics. This is the kind of energy that is generated as the result of any thermal activity, at the lower thermal level, and is not utilized by the process.

It was in 1872, though, that Boltzmann used this concept to create his *theory of statistical thermodynamics*, thus expressing the uncertainty of the state of the molecules of a perfect gas. The idea was created by the inability of the dynamic theory to account for all the collisions of the molecules, which generate the thermal energy. Boltzmann stated that the entropy of a perfect gas, changing states isothermally, at temperature T is given by

$$S = -k\int_x (\psi - H)/kT \exp\{(\psi - H)/kT\}\, dx, \tag{6.1}$$

where ψ is the Gibbs energy, $\psi \doteq -kT \ln \exp\{-H/kT\}$, H is the total energy of the system, and k is Boltzmann's universal constant. Due to the size of the problem and the uncertainties involved in describing its dynamic behavior, a probabilistic model was assumed where the entropy is a measure of the molecular distribution. $p(x)$ is defined as the probability of a molecule being in state x, thus assuming that

$$p(x) = \exp\{(\psi - H)/kT\}, \tag{6.2}$$

where $p(x)$ must satisfy the incompressibility property over time, of the probabilities, in the state space X, for example,

$$dp/dt = 0. \tag{6.3}$$

The incompressibility property is a differential constraint when the states are defined in a continuum, which in the case of perfect gases yields the Liouville equation. Substituting Eq. (6.2) into Eq. (6.1), the entropy of the system takes the form

$$S = -k \int_x p(x) \ln p(x) \, dx. \tag{6.4}$$

The above equation defines entropy as a measure of the uncertainty about the state of the system, expressed by the probability density exponential function of the associated energy.

Actually, the problem of describing the entropy of an isothermal process should be derived from the dynamical theory of thermodynamics, considering heat as the result of the kinetic and potential energies of molecular motion. It is the analogy of the two formulations that led to the study of the equivalence of entropy with the performance measure of a control system. If the dynamical theory of thermodynamics is applied on the aggregate of the molecules of a perfect gas, an average Lagrangian I, should be defined to describe the average performance over time of the state x of the gas,

$$I = \int_{t_0}^{t_f} L(x, t) \, dt, \tag{6.5}$$

where the Lagrangian $L(x, t) =$ (kinetic energy) $-$ (potential energy). The average Lagrangian when minimized, satisfies the second law of thermodynamics. Because the formulations of Eqs. (6.1) and (6.5) are equivalent, the following relation should be true:

$$S = I/T, \tag{6.6}$$

where T is the constant temperature of the isothermal process of a perfect gas [Lindsay and Margenau (1957)]. This relation will be the key in order to express the performance measure of the control problem as entropy.

6.1.2 Entropy and Information Theory

In the 1940s Shannon (1963), using Boltzmann's idea, for example, Eq. (6.4), defined entropy (negative) as a measure of the uncertainty of the transmission of information, in his celebrated work on *information theory*:

$$H = -\int_{\Omega} p(s) \ln p(s) \, ds, \tag{6.7}$$

where $p(s)$ is a Gaussian density function over the space Ω of the information signals transmitted. The similarity of the two formulations is obvious, where the uncertainty about the state of the system is expressed by an exponential density function of the energy involved.

Shannon's theory was generalized for dynamic systems by Ashby (1975), Boettcher and Levis (1983), and Conant (1976), who also introduced various laws that cover information systems, like the partition law of information rates.

6.1.3 Jaynes's Principle of Maximum Entropy

In an attempt to generalize the principle used by Boltzmann and Shannon to describe the uncertainty of the performance of a system under a certain operating condition, Jaynes (1957) formulated his *maximum entropy principle*, to apply it in theoretical mechanics. In summary it claims that

> the uncertainty of an unspecified relation of the function of a system is expressed by an exponential density function of a known energy relation associated with the system.

A modified version of the principle, as it applies to the control problem, will be derived in the following using Calculus of Variations [Saridis, (1988)]. The proposed derivation represents a new formulation of the control problem, either for deterministic or stochastic systems and for optimal or nonoptimal solutions.

6.1.4 Uncertainty and the Control Problem

Uncertainty, thus far, has been associated with insufficient knowledge and lack of information in thermodynamics and information theory. The models used were probabilistic. As such, one may distinguish two categories:

- Subjective (reducible) uncertainties
- Objective (irreducible) uncertainties

The first category refers to the lack of knowledge of the experimenter, and the latter refers to phenomena from which no further improvement of knowledge is attainable.

Now the control design problem will be associated with subjective uncertainty. The designer is faced with the uncertainty of optimal design before solving the problem. He may assign an appropriate probability density function over the space of *admissible controls*, and reduce the design problem into the problem of finding the point (control law), that has *maximum probability of attaining the optimal value*. This approach will be pursued in the sequel with *entropy*, the measure of uncertainty.

Other attempts have also been made to formulate the estimation problem [Kalata and Premier (1974)], and the feedback control problem [Weidemann

(1969)], using entropy methods. However, those attempts found little success in the literature.

6.2 MAXIMUM ENTROPY FORMULATION OF OPTIMAL AND SUBOPTIMAL CONTROL

The purpose of this work is to establish entropy measures, equivalent to the performance criteria of the optimal control problem, while providing a physical meaning to the latter. This is done by expressing the problem of control system design probabilistically and assigning a distribution function representing the uncertainty of selection of the optimal solution over the space of admissible controls. By selecting the worst-case distribution, satisfying *Jaynes' maximum entropy principle*, the performance criterion of the control is associated with the entropy of selecting a certain control [Jaynes (1957), Saridis (1985b)]. Minimization of the differential entropy, which is equivalent to the average performance of the system, yields the optimal control solution. Furthermore, the generalized Hamilton–Jacobi–Bellman equation is derived from the incompressibility over time condition of the probability distribution. Adaptive control and stochastic optimal control are obtained as special cases of the optimal formulation, with the differential entropy of active transmission of information, claimed by Fel'dbaum (1965), as their difference. Upper bounds of the latter may yield measures of goodness of the various stochastic and adaptive control algorithms.

In this section, the entropy measure for optimal control will be established.

6.2.1 Deterministic Optimal Control Problem

The optimal feedback deterministic control problem with accessible states is defined as follows: Given the dynamic system

$$dx/dt = f(x,u,t), \qquad x(t_0) = x_0, \tag{6.8}$$

and the cost function,

$$V(u; x_0, t_0) = \int_{t_0}^{T} L(x,u,t)\, dt, \tag{6.9}$$

where $x(t) \in \Omega_x$ is the n-dimensional state vector, $u(x,t) \in \Omega_u \times T \subset \Omega_x \times T$, is the m-dimensional feedback control law, and $t \in \Im = [t_0, T]$.

An optimal control $u^*(x,t)$ is sought to minimize the cost:

$$V(u^*; x_0, t_0) = \min_u \int_{t_0}^{T} L(x,u,t)\, dt. \tag{6.10}$$

Define the differential entropy, for some $u(x,t)$,

$$H(x_0, u(x,t), p(u)) = H(u) = -\int_{\Omega_{x_0}}\int_{\Omega_x} p(x_0, u)\ln p(x_0, u)\, du\, dx_0, \tag{6.11}$$

where $x_0 \in \Omega_{x_0}$, $x \in \Omega_x$ are the spaces of initial conditions and states, respectively, and $p(x_0, u) = p(u)$ is the probability density of selecting u. One may select the density function $p(u)$ to maximize the differential entropy according to Jaynes' maximum entropy principle, subject to $E\{V(x_0, u, t)\} = K$, for some $u(x,t)$. This represents a problem more general than the optimal, where K is a fixed but unknown constant, depending on the selection of $u(x,t)$. The unconstrained expression of the differential entropy is

$$\begin{aligned} I &= \beta H(u) - \gamma[E\{V\} - K] - \alpha\left[\int_{\Omega_x} p(u)\, dx - 1\right] \\ &= -\int_{\Omega_x}[\beta p(u)\ln p(u) + \gamma p(u)V]\, dx - \alpha\left[\int_{\Omega_x} p(u)\, dx - 1\right]. \end{aligned} \tag{6.12}$$

Using the lemma of calculus of variations, maximization of I with respect to $p(u)$ yields

$$\partial/\partial[-\beta p(u)\ln p(u) - \gamma p(u)V - \alpha p(u)] = 0, \qquad \partial^2 I/\partial p^2 < 0,$$

or

$$-\beta \ln p(u) - \beta - \gamma V - \alpha = 0, \qquad -\beta/p(u) < 0,$$

and the worst-case density is

$$\begin{aligned} p(u) &= e^{-\lambda - \mu V(u(x,t),\, x0,\, t0)} \\ e^{\lambda} &= \int_{\Omega_x} e^{-\mu V(u(x,t),\, x0,\, t0)}\, dx, \end{aligned} \tag{6.13}$$

with $\mu = \gamma/\beta$, $\lambda = (\alpha + 1)/\beta$, and

$$E\{V(u(x,t), x_0, t_0)\} = K = \partial\left[\ln \int_{\Omega_x} e^{-\mu V}\, dx\right]\bigg/\partial\mu.$$

Now let us recast the optimal feedback control problem as the optimal design problem under the worst uncertainty of selecting the best control, from the set of admissible feedback controls, $u(x,t) \in \Omega_u \times T \subset \Omega_x \times T$. It is assumed that the set of admissible controls is covered by the probability density function $p(u)$, given by Eq. (6.13), expressing the uncertainty of selection of the optimal function. The worst-case (maximum) differential entropy corresponding to $p(u)$ is given by

$$H(u) = \lambda + \mu E\{V(u(x,t), x_0, t_0)\}, \tag{6.14}$$

and the corresponding minimum value with respect to $u(x,t)$ represents the optimal design. This condition implies

$$dH/du = \partial H/\partial u + \partial H/\partial p, \qquad \partial p/\partial u = 0 \qquad \text{(Frechet derivative)}. \tag{6.15}$$

However, Jaynes's principle implies

$$\partial H/\partial p = 0.$$

Therefore,

$$dH/du = \partial H/\partial u. \tag{6.16}$$

The following theorem establishes the equivalence of the optimal feedback control problem with the minimization of the worst-case differential entropy.

Theorem 6.1 A necessary and sufficient condition for $u^*(x,t)$ to minimize $V(u(x,t), x_0, t_0)$, subject to the system equations,

$$dx/dt = f(x,u,t), \qquad x(t_0) = x_0,$$

is the $u^*(x,t)$ that minimizes the differential entropy $H(u, p(u))$, where $p(u)$ is the worst entropy density function according to Jaynes' maximum entropy principle.

The proof of the theorem is given in Saridis (1988). Furthermore, the deterministic version of the approximation theory of nonlinear optimal control [Saridis and Lee (1979)] may be obtained from the properties of the density function $p(u)$:

$$p(u) = e^{-\lambda - \mu V(u(x,t), x_0, t_0)}$$

$$e^{\lambda} = \int_{\Omega_x} e^{-\mu V(u(x,t), x_0, t_0)}\, dx. \tag{6.17}$$

This density must satisfy the time incompressibility condition of probability functions, equivalent to the Liouville equation [Prigogine (1980)]:

$$dp/dt = 0 \Leftrightarrow \left[\partial p/\partial t + \partial p^T/\partial x \cdot dx/dt + L(x,u,t)\right] p = 0, \tag{6.18}$$

where

$$\partial p/\partial t = -\partial V/\partial t \cdot p(u), \quad \partial p(u)/\partial x = -\partial V/\partial x \cdot p(u),$$

and

$$dx/dt = f(x,u,t)$$

for some given $u(x,t)$. Then,

$$\left[\partial V/\partial t + \partial V^T/\partial x \cdot f(x,u,t) + L(x,u,t)\right] p(u) = 0.$$

Because Eq. (6.13) implies that $p(u) \neq 0$, for all $x \in \Omega_x$;

$$\left[\partial V/\partial t + \partial V^T/\partial x \cdot f(x,u,t) + L(x,u,t)\right] p = 0, \qquad V(X(T),T) = 0, \tag{6.19}$$

which is the generalized Hamilton–Jacobi equation for an arbitrary admissible control, derived by Saridis and Lee (1979). Minimization of Eq. (6.19) yields the optimal control that minimizes the differential entropy $H(u)$; for example,

$$\partial V/\partial t + \min_u \left[\partial V^T/\partial x \cdot f(x,u,t) + L(x,u,t)\right] p = 0, \qquad V(x(T),T) = 0. \tag{6.20}$$

The above formulation was extended by Loparo and his colleagues to the deterministic discrete-time case in Tsai, Casiello, and Loparo (1992).

6.2.2 Stochastic Optimal-Control Problem

The stochastic version of the theorem, which is appropriate for the present formulation, is obtained the same way as the deterministic case. Assume the following stochastic system and performance cost are given. The system equations are

$$\begin{aligned} dx/dt &= f[x,u,t] + g(x,t)w(t), \qquad x(t_0) = x_0, \qquad p(x_0) \\ z(t) &= h(x,t) + v(t), \end{aligned} \tag{6.21}$$

where the random processes and their associated probability densities, $w(t)$, $p(w)$: $v(t)$, $p(v)$, are given.

The performance cost is

$$\begin{aligned} V[u(t), x_0, t_0] &= E\{J[x_0, t_0, u(x,t), w(t), v(t)]\} \\ &= E\left\{\int_{t_0}^{T} L(x,u,t)\, dt/x_0\right\}. \end{aligned} \tag{6.22}$$

Derive $p(u(x,t))$ as Jaynes's worst entropy probability density function:

$$p(u(x,t)) = \exp[-\lambda - \mu E\{V[u(x,t), x_0, t_0]\}]. \qquad (6.23)$$

Then the associated entropy is

$$H[u(x,t)] = \lambda + \mu E\{V[u(x), x_0, t_0]\}, \qquad (6.24)$$

which may replace the performance criterion for optimization. Jaynes's principle of maximum entropy implies

$$dH/dp = 0.$$

The incompressibility in time of probability density condition implies

$$dp/dt = 0,$$

which, for the selected $p(u(x,t))$ yields the *generalized Hamilton–Jacobi–Bellman* equation, used in the previous chapter for the approximation theory of stochastic systems:

$$\begin{aligned} &\partial V/\partial t + \partial V^T/\partial x \cdot f(x,u,t) + L(x,u,t) \\ &\quad + \tfrac{1}{2}\pounds\{\mathrm{Tr}(\partial V^2/\partial x^2\, Qgg^T)\} = 0, \qquad V(T) = 0. \end{aligned} \qquad (6.25)$$

Minimization of Eq. (6.25) yields again the stochastic optimal control

$$\begin{aligned} &\partial V/\partial t + \min_u \big[\partial V^T/\partial x \cdot f(x,u,t) + L(x,u,t) \\ &\quad + \tfrac{1}{2}\pounds\{\mathrm{Tr}(\partial V^2/\partial x^2 Qgg^T)\}\big] = 0, \qquad V(x(T),T) = 0. \end{aligned} \qquad (6.26)$$

Theorem 6.2 A necessary and sufficient condition for $u(x,t)$ to minimize

$$V[u(x,t), x_0, t_0] = E\{J(u(x,t), x_0, t_0)\}$$

subject to the dynamic constraints:

$$\begin{aligned} dx/dt &= f(x,u,t) + g(x,t)w(t), \qquad x(t_0) = x_0, \\ z(t) &= h(x,t) + v(t), \end{aligned}$$

where $w(t), v(t)$, are i.i.d. random processes, is that $u(x,t)$ minimizes the *differential entropy* $H[u(x,t)]$, where the associated $p(u(x,t))$ is the *maximum entropy density function*, satisfying *Jaynes' maximum entropy principle*.

The above theorem may be used to obtain the solution of the stochastic optimal-control problem. It may also be extended to define other design problems as minimum entropy problems, for example, the parameter identification, optimal state estimation and dual control problems. It also serves in establishing analytically Fel'dbaum's claim that the *feedback active optimal control* is really the globally minimal solution of the stochastic optimal control problem.

6.2.3 Stochastic Suboptimal-Control Problem

The suboptimal-control problem resulting from the approximation theory developed by Saridis and Lee (1979), and in Chapter 5, may be reformulated using Jaynes's principle of maximum entropy.

> The approximation theory for stochastic systems claims that the performance cost function $V(u(x); x, t)$ satisfies the generalized Hamilton–Jacobi–Bellman equation for an admissible control $u(x) = u_i(x)$, $i = 1, 2, \ldots$.

Such a generalized Hamilton–Jacobi–Bellman equation was derived in (6.25), for some control $u(x)$, using Jaynes's principle of maximum entropy. To complete the claim one may assume that $u(x) = u_i(x)$ in Eq. (6.25). Then the following theorem establishes the claim.

Theorem 6.3 The performance cost $V(u_i(x); x, t)$, of system (6.14) for an admissible control $u_i(x)$, may be expressed by its associated entropy $H(u_i(x, t)$, $i = 1, \ldots$ derived by Jaynes's principle of maximum entropy:

$$u_i(x) = -\tfrac{1}{2}b^T \partial V_{i-1}/\partial x \Rightarrow V_{i-1} \geq V_i.$$

The derivations of the rest of the theorems of the approximation theory of stochastic systems follow the same procedure as in Chapter 5. Their proofs are presented in Saridis (1988). This includes the upper and lower bounds of the approximation, the infinite-time optimal-control problem, and the design procedures. The optimal deterministic and stochastic control are special cases of the above theorem.

6.2.4 Discrete-Time Formulation of the Stochastic Optimal-Control Problem

The discrete-time stochastic optimal-control problem may be reformulated as an entropy minimization problem, associated with the worst-case probability, in a way similar to the continuous-time case presented in Section 6.2.2. The entropy approach to include both deterministic and stochastic discrete-time optimal-control problems was presented in Tsai, Casiello, and Loparo (1992). The stochastic version of the entropy formulation of the discrete-time stochastic optimal-control problem is presented here. Assume that the following discrete-time stochastic system, performance cost, random processes

$w(k), v(k)$, and their respective associated probability densities $p(w), p(v)$ are given.

The system equations are

$$x(k+1) = f(x(k), u(k), k) + g(x(k), k)w(k), \qquad x(0) = x_0, p(x_0)$$
$$z(k) = h(x(k), k) + v(k), \tag{6.27}$$

where $u(k) = u(x, k) \in \Omega_u$; Ω_u is the set of admissible controls.

The performance cost is

$$V_N[u(x,k), x_0] = E\{J[x_0, u(x,k), w(k), v(k)]/x_0\}$$
$$= E\left\{\sum_{k=0}^{N-1} L(x(k+1), u(k), k)/x_0\right\}. \tag{6.28}$$

Derive $p(u(x,k))$, as Jaynes's worst entropy probability density function:

$$p(u(x,k)) = \exp[-\lambda - \mu E\{V[u(x,k), x_0]\}]. \tag{6.29}$$

Then the associated entropy is

$$H[u(x,k)] = -\int_{\Omega_x} p(u(x)) \ln p(u(x))\, dx = \lambda + \mu E\{V[u(x,k), x_0]\}, \tag{6.30}$$

which may replace the performance criterion for optimization.

The worst-case density function according to Jaynes's principle of maximum entropy was obtained from

$$dH/dp = 0. \tag{6.31}$$

The Bellman equation that governs the dynamics of the discrete-time system is given, for an arbitrary but fixed admissible controller $\{u_i(k)\}$ used in the approximation theory of Chapter 5, by

$$V_{N-k}(x(k)/Z^k)$$
$$= E\{L(x(k+1), u_i(k)) + V_{N-k-1}(x(k+1)/Z^{k+1})/Z^k\}, \tag{6.32}$$

and for the optimal controller $u^*(k)$,

$$V_{N-k}(x(k)/Z^k)$$
$$= \min_u E\{L(x(k+1), u_i(k)) + V_{N-k-1}(x(k+1)/Z^{k+1})/Z^k\}. \tag{6.33}$$

Minimization of Eq. (6.30) yields again the stochastic optimal control for the discrete-time problem.

Theorem 6.4 A necessary and sufficient condition for $u(x,t)$ to minimize

$$V[u(x,k),x_0] = E\{J(u(x,k),x_0)\}$$

subject to the dynamic constraints

$$x(k+1) = f(x(k),u(k),k) + g(x(k),k)w(k), \qquad x(0) = x_0, p(x_0),$$

$$z(k) = h(x(k),k) + v(k),$$

where $w(k), v(k)$ are i.i.d. random processes, is that $u(x,k)$ minimizes the *differential entropy* $H[u(x,k)]$, where the associated $p(u(x,k))$ is the *maximum entropy density function*, satisfying *Jaynes' maximum entropy principle.*

The above theorem may be used to obtain the solution of the discrete-time stochastic optimal-control problem. It may also be extended to define other design problems as minimum entropy problems, for example, the parameter identification, optimal state estimation, and dual control problems. It also serves in establishing analytically Fel'dbaum's claim that the *feedback active optimal control* is really the globally minimal solution of the stochastic optimal-control problem. The state-estimation problem will be treated in the following.

6.3 MAXIMUM ENTROPY FORMULATION OF STATE ESTIMATION

State estimation for continuous-time and discrete-time systems can be easily formulated using entropy measures and Jaynes's principle of maximum entropy.

6.3.1 Continuous-Time State Estimation

The state-estimation problem both in the linear case, for example, the Kalman–Bucy filter, as well as the nonlinear case, for example, the MMSE estimator, may be reformulated as a maximum entropy problem by using Jaynes's principle of *maximum entropy*. Following a procedure similar to the optimal-control reformulation for the continuous-time system, Eq. (6.21),

$$dx/dt = f(x,u,t) + g(x,t)w(t), \qquad x(t_0) = x_0, p(x_0)$$

$$z(t) = h(x,t) + v(t), \qquad w(t); p(w): v(t); p(v), \tag{6.34}$$

and the risk function, for $t \in \mathfrak{I} = [t_0, T]$,

$$R'(\hat{x}) = E\left\{\int_{\mathfrak{I}} (\hat{x} - x)^T A(\hat{x} - x)\, dt\right\}, \tag{6.35}$$

where $A = \Lambda^{-1}$, the inverse of the measurement noise covariance matrix. It is easy to find the worst-case entropy according to Jaynes's principle:

$$\begin{aligned} I' &= \beta H(\hat{x}) - \gamma\left[E\left\{\int_{\mathfrak{I}} \|\hat{x} - x\|_A^2\, dt\right\} - K\right] - \alpha\left[\int_{\Omega_x} p(\hat{x})\, d\hat{x} - 1\right] \\ &= -\int_{\Omega_{\hat{x}}}\left[\beta p(\hat{x}) \ln p(\hat{x}) + \gamma p(\hat{x}) \int_{\mathfrak{I}} \|\hat{x} - x\|_A^2\, dt\right] d\hat{x} - \alpha\left[\int_{\Omega_x} p(\hat{x})\, d\hat{x} - 1\right]. \end{aligned} \tag{6.36}$$

Using the lemma of calculus of variations, maximization of I with respect to $p(u)$ yields

$$\partial\left[-\beta p(\hat{x}) \ln p(\hat{x}) - \gamma p(\hat{x}) \int_{\mathfrak{I}} \|\hat{x} - x\|_A^2\, dt - \alpha p(\hat{x})\right] \Big/ \partial p = 0,$$

$$\partial^2 I / \partial p^2 < 0$$

or

$$-\beta \ln p(\hat{x}) - \beta - \gamma \int_{\mathfrak{I}} \|\hat{x} - x\|_A^2\, dt - \alpha = 0, \qquad -\beta / p(\hat{x}) < 0,$$

and the worst-case density is

$$\begin{aligned} p(\hat{x}) &= \exp\left\{-\upsilon - \rho \int_{\mathfrak{I}} \|\hat{x} - x\|_A^2\, dt\right\}, \\ e^{\lambda} &= \int_{\Omega_{\hat{x}}} \exp\left\{-\rho \int_{\mathfrak{I}} \|\hat{x} - x\|_A^2\, dt\right\} dx, \end{aligned} \tag{6.37}$$

with $\upsilon = \gamma/\beta$, $\rho = (\alpha + 1)/\beta$, and

$$R'(\hat{x}) = K = \partial/\partial \upsilon \ln \int_{\Omega_{\hat{x}}} \exp\left\{-\upsilon \int_{\mathfrak{I}} \|\hat{x} - x\|_A^2\, dt\right\} dx \tag{6.38}$$

The associated worst-case entropy then is given by

$$H(\hat{x}) = \rho + \upsilon E\left\{\int_{\mathfrak{I}} [\hat{x} - x]^T \Lambda^{-1} [\hat{x} - x]\, dt\right\}, \tag{6.39}$$

and minimization of $H(\hat{x})$ yields the optimal MMSE (minimum mean-square-error) estimator. The following theorem is a result of the previous formulation.

Theorem 6.5 The optimal MMSE state estimator for the system of Eq. (6.35) is obtained by minimizing the differential entropy $H(\hat{x})$, where the associated $p(\hat{x})$ is the *maximum entropy density function* satisfying *Jaynes' maximum entropy principle.*

In the case of a linear Gaussian system, the MMSE estimator gives the most probable estimator as well, according to the Bayes theorem. This problem will be considered in the following as part of the entropy formulation of the linear-quadratic-gaussian (LQG) optimal control problem. Alternate formulations of the optimal state-estimation problem using entropy criteria are also available in the literature [Kalata and Premier (1974)].

6.3.2 Discrete-Time State Estimation

The problem of discrete-time state estimation is also a direct extension of the continuous-time estimation problem discussed in the previous section. This problem was also treated by Loparo and his colleagues in Tsai, Casiello, and Loparo (1992). The results of the entropy formulation are presented hereafter.

Assume that the following discrete-time stochastic system, performance cost, random processes $w(k)$, $v(k)$, and their respective associated probability densities $p(w)$, $p(v)$ are given. The system equations are

$$\begin{aligned} x(k+1) &= f[x(k), u(k), k] + g[x(k), k]w(k), \qquad x(0) = x_0, p(x_0), \\ z(k) &= h[x(k), k] + v(k), \end{aligned} \tag{6.40}$$

where $u(k) = u(x, k) \in \Omega_u$, Ω_u is the set of admissible controls, and the risk function, for $k \in \mathscr{H} = [0, N]$, is

$$R'(\hat{x}) = E\left\{ \sum_{k=0}^{N-1} [\hat{x}(k) - x(k)]^T \Lambda^{-1} [\hat{x}(k) - x(k)] \Big/ Z^{N-1} \right\}. \tag{6.41}$$

The worst-case entropy according to Jaynes's principle is found in a similar way as previously, and the worst-case density function is

$$\begin{aligned} p(\hat{x}) &= \exp\left[-\lambda - \mu E\left\{ \sum_{k=0}^{N-1} [\hat{x}(k) - x(k)]^T \Lambda^{-1} [\hat{x}(k) - x(k)] \Big/ Z^{N-1} \right\} \right], \\ e^{\lambda} &= \int_{\Omega_{\hat{x}}} \exp\left\{ -\rho \sum_{k=0}^{N-1} \| \hat{x} - x \|_A^2 \right\} dx, \end{aligned} \tag{6.42}$$

and

$$R'(\hat{x}) = \partial/\partial v \ln \int_{\Omega_{\hat{x}}} \exp\left\{-v \sum_{k=0}^{N-1} \|\hat{x} - x\|_A^2\right\} dx. \tag{6.43}$$

The associated worst-case entropy then is given by

$$H(\hat{x}) = \rho + vE\left\{\sum_{k=0}^{N-1} [\hat{x}(k) - x(k)]^T \Lambda^{-1} [\hat{x}(k) - x(k)]/Z^{N-1}\right\}, \tag{6.44}$$

and minimization of $H(\hat{x})$ yields the optimal MMSE (minimum risk) estimator. The following theorem is a result of the above formulation.

Theorem 6.6 The optimal MMSE state estimator for the system of Eq. (6.33) is obtained by minimizing the differential entropy $H(\hat{x})$, where the associated $p(\hat{x})$ is the *maximum entropy density function* satisfying *Jaynes' maximum entropy principle*.

In the case of a linear Gaussian system, the MMSE estimator gives the most probable estimator as well, according to the Bayes theorem. This problem will be considered in the sequel as part of the entropy formulation of the linear-quadratic-Gaussian (LQG) optimal-control problem.

6.4 COST OF ACTIVE FEEDBACK (DUAL) CONTROL PROBLEM

In Chapter 4, the case of active feedback (dual) control was presented with no mathematical verification, due to the lack of analytic tools. The entropy reformulation of the stochastic optimal-control problem provides those tools through the partition properties of the differential entropy of the control and state-estimation activities. The cases of stochastic optimal control and stochastic optimal adaptive control will be considered here, and thus analytically verify Fel'dbaum's claim that only the *active feedback optimal control* is the globally optimal solution.

Most of the existing stochastic Open-Loop-Feedback-Optimal and Adaptive control algorithms have been derived based on the *certainty equivalence principle* [Kumar and Varaiya (1986)]:

> The stochastic optimal or adaptive control has the same form as if the estimate of the state and/or the unknown parameter were certain, without error.

This principle is correct only for LQG problems discussed in Chapter 4, but is also useful in designing approximate controls when the estimation process converges fast enough. In this section, it will be shown through the entropy partition theorem that such solutions are not optimal, but one may establish measures of deviation from optimality.

6.4.1 Stochastic Optimal Control

Consider the worst-case differential entropy of the optimal control $H(u^*)$. It may be partitioned as follows, using the definition of marginal probabilities, to take into account the underlying state-estimation problem:

$$\begin{aligned}
H(u^*) &= -\int_{\Omega_x} p(u^*) \ln p(u^*)\, dx = -\int_{\Omega_x}\int_{\Omega_{\hat{x}}} p(u^*, x) \ln p(u^*)\, d\hat{x}\, dx \\
&= -\int_{\Omega_x}\int_{\Omega_{\hat{x}}} p(u^*/\hat{x}) p(\hat{x}) \ln[p(u^*/\hat{x}) p(\hat{x})/p(\hat{x}/u^*)]\, d\hat{x}\, dx \\
&= -\int_{\Omega_x}\left[\int_{\Omega_{\hat{x}}} p(u^*/\hat{x}) p(\hat{x}) \ln[p(u^*/\hat{x})]\, dx\right] p(x)\, d\hat{x} \\
&\quad -\int_{\Omega_x}\left[\int_{\Omega_{\hat{x}}} p(u^*/\hat{x})\, dx\right] \ln p(\hat{x})\, d\hat{x} \\
&\quad -\int_{\Omega_x}\left[\int_{\Omega_{\hat{x}}} p(\hat{x}/u^*) \ln p(\hat{x}/u^*)\, d\hat{x}\right] p(u^*)\, dx.
\end{aligned}$$

The final expression may be rewritten in terms of differential entropies as

$$H[u^*] = H[u^*/\hat{x}] + H[\hat{x}] - H[\hat{x}/u^*]. \tag{6.45}$$

The decomposition (6.45), explained by the following theorem, justifies Fel'dbaum's claim that the active feedback control accounting for the control function when estimating future states is *globally minimum*. It also disclaims that the partition theorem of stochastic control is optimal in all cases outside the LQG.

Theorem 6.7 The cost of optimal control with inaccessible states is the cost of control with estimates of the states x, plus the cost of $\hat{x}$, plus the cost of state estimation, minus the cost of active transmission of information.

6.4.2 Stochastic Suboptimal Control

A similar decomposition and a suitable theorem applies to the suboptimal control of Section 6.2.3:

$$H[u_i] = H[u_i/\hat{x}] + H[\hat{x}] - H[\hat{x}/u_i]. \tag{6.46}$$

Theorem 6.8 The cost of suboptimal control u_i with inaccessible states is the cost of control with estimates of the states x, plus the cost of $\hat{x}$, plus the cost of state estimation, minus the cost of active transmission of information for control u_i.

6.4.3 Stochastic Optimal Adaptive Control

The *optimal adaptive control* problem evolves from the stochastic optimal-control problem when the system parameters $\phi \in \Omega_\phi$, are unknown or unaccessible for measurement. The system may be described by

$$dx/dt = f(x,u,\phi,t) + g(x,\phi,t)w(t), \qquad x(t_0) = x_0; p(x_0),$$
$$z(t) = h(x,t) + v(t), \qquad w(t); p(w): v(t); p(v), \tag{6.47}$$

and it is required that

$$H(u^*) = \min_u H(u) \Leftrightarrow \min_u (x_0, u, t), \tag{6.48}$$

subject to optimal state estimation and parameter identification. Assuming a least-squares algorithm for parameter identification ϕ, a partition of the differential entropy of the stochastic system $H(u^*)$, similar to the *stochastic optimal-control* problem, yields

$$H[u^*] = H\left[u^*/\hat{x},\hat{\phi}\right] + H\left[\hat{x}/\hat{\phi}\right] + H\left[\hat{\phi}\right] - H\left[\hat{x},\hat{\phi}/u^*\right], \tag{6.49}$$

where $H[\hat{x},\hat{\phi}/u^*]$ is the active transmission of information both in state estimation $\hat{x}$, and parameter identification $\hat{\phi}$.

A theorem similar to 6.7 may be stated for the case of the optimal adaptive control.

Theorem 6.9 The cost of control with inaccessible states and parameters is the cost of control with estimates of the states $\hat{x}$ and parameters $\hat{\phi}$, plus the cost of state estimation with estimates of parameters $\hat{\phi}$, minus the cost of active transmission of information of the states and parameters.

Typical Adaptive Control algorithms [Saridis (1977), Kumar and Varaiya (1986), Morse (1992)] are based on the *certainty equivalence principle*, and take into consideration only the cost control using the state and parameter estimates and that of the state estimation and parameter identification. It is clear from the above formulation that these algorithms are *definitely not optimal*.

This is demonstrated by the following inequality:

$$\begin{aligned} H[u]_{\text{ad}} &= H\left[u^*/\hat{x},\hat{\phi}\right] + H\left[\hat{x}/\hat{\phi}\right] + H\left[\hat{\phi}\right] \\ &= H(u^*) + H\left[\hat{x},\hat{\phi}/u^*\right] > H(u^*). \end{aligned} \tag{6.50}$$

The following example demonstrates this statement, which agrees with Fel'dbaum's claim of dual control.

EXAMPLE: **The Dual-Optimal and Adaptive Control** The difference between the dual-optimal and adaptive control is demonstrated with the comparison of an adaptive to an LQG optimal-control problem. Given the system

$$\begin{aligned} dx_1/dt &= x_2, & x(0) &\sim N(0,10I), \\ dx_2/dt &= \phi x_2 + u + w, & \phi &= -1, \\ z &= x_1 + v, & w(t) &\sim (0,1), \qquad v(t) \sim (0,1), \end{aligned}$$

and the performance cost

$$J(u) = E\left\{\int_0^T \left(2x_1^2 + 2x_2^2 + \tfrac{1}{2}u^2\right) dt\right\}.$$

The problem is to calculate, for the cost function J and ϕ unknown (a) the adaptive solution, obtained from the principle certainty equivalence, and (b) the case that ϕ is known, for example, the optimal solution.

(a) The Adaptive Case

$$\begin{aligned} H(u^*,\hat{x},\hat{\phi}) &= J\left(x(0),u^*(\hat{x},\hat{\phi})\right) \\ &= E\Big\{\left[9+\hat{\phi}+\hat{\phi}^2+(\hat{\phi}+1)\sqrt{\hat{\phi}^2+8}\right] \\ &\qquad \times x(0)_1^2\Big/\left[(\hat{\phi}+1)+\sqrt{\hat{\phi}^2+8}\right] + 2x(0)_1 x(0)_2 \\ &\qquad + \left[4+\tfrac{1}{2}\hat{\phi}^2+\tfrac{1}{2}\hat{\phi}\sqrt{\hat{\phi}^2+8}\right]x(0)_2^2\Big/\left[(\hat{\phi}+1)+\sqrt{\hat{\phi}^2+8}\right]\Big\}, \end{aligned}$$

with

$$u^*(x) = -2\hat{x}_1 - \left[\hat{\phi}+\sqrt{\hat{\phi}^2+8}\right]x(0)_2,$$

$$H(\hat{x}/\hat{\phi}) = \int_0^T \operatorname{Tr} P(t)\, dt = \int_0^T \left(p(t)_{11} + p(t)_{22}\right) dt,$$

where

$$P(t) = \begin{bmatrix} p_{11} & p_{12} \\ p_{12} & p_{22} \end{bmatrix},$$

$P(0) = 10I$, the error covariance matrix and

$$\begin{aligned} dp_{11}/dt &= 2p_{12} - p_{11}^2, & p(0)_{11} &= 10, \\ dp_{12}/dt &= p_{22} + \phi p_{12} - p_{11}p_{12}, & p(0)_{22} &= 10, \\ dp_{22}/dt &= 2\phi p_{22} - p_{12}^2, & p(0)_{12} &= 0, \end{aligned}$$

and x is estimated by a Kalman filter,

$$\begin{aligned} d\hat{x}_1/dt &= \hat{x}_2 + p_{11}(z - \hat{x}_1), \\ d\hat{x}_2/dt &= \hat{\phi}\hat{x}_1 + p_{12}(z - \hat{x}_1). \end{aligned}$$

The parameter estimation is performed with a least-squares identification algorithm [Saridis (1977)], after time discretization:

$$\begin{aligned} \hat{f}(k+1) &= \hat{f}(k) + S(k+1)\omega(k)\big[\omega(k+1) - T^2\omega(k)\big], \\ S(k+1) &= S(k) - S(k)^2\omega(k)^2[\omega(k)S(k)\omega(k) + 1]^{-1}, \\ \omega(k) &= z(k) - z(k-1), \\ \hat{f}(k) &= 1 + \hat{\phi}T; \qquad T = 0.1 \text{ s}. \end{aligned}$$

The results of the state estimation are given in Figure 6.1, and the results of the parameter identification in Figure 6.2. The latter have been used in the state-estimation algorithm. The controller $u^*(\hat{x}, \hat{\phi})$ uses the results of both estimation and identification algorithms and is depicted in Figure 6.3.

(b) The Optimal LQG Case. The optimal LQG solution is obtained by the following steps

$$H(u/\hat{x}) = J(x_0, u(\hat{x})) = E\left\{\int_0^T \left(2\hat{x}_1^2 + 2\hat{x}_2^2 + \tfrac{1}{2}u^2\right) dt\right\},$$

and the optimal controller is

$$u^*(\hat{x}) = -2\hat{x}_1 - 2\hat{x}_2.$$

The differential entropy of estimation $H(\hat{x})$ is obtained from the state

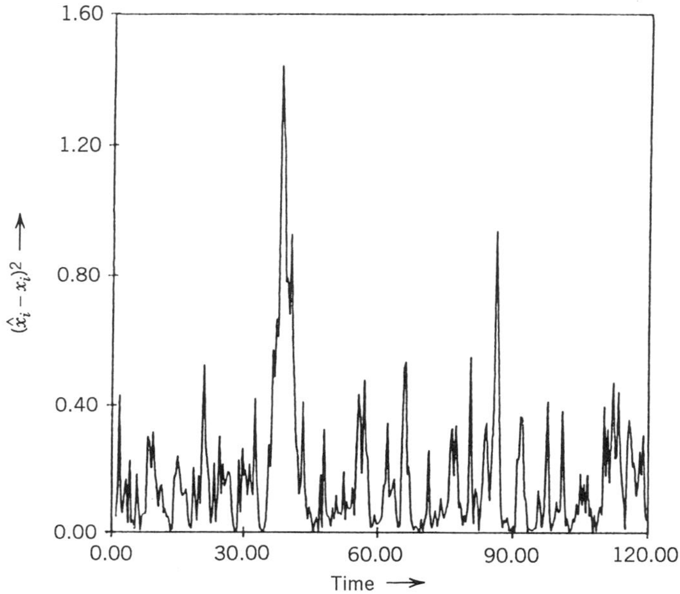

Fig. 6.1 Sum of square errors

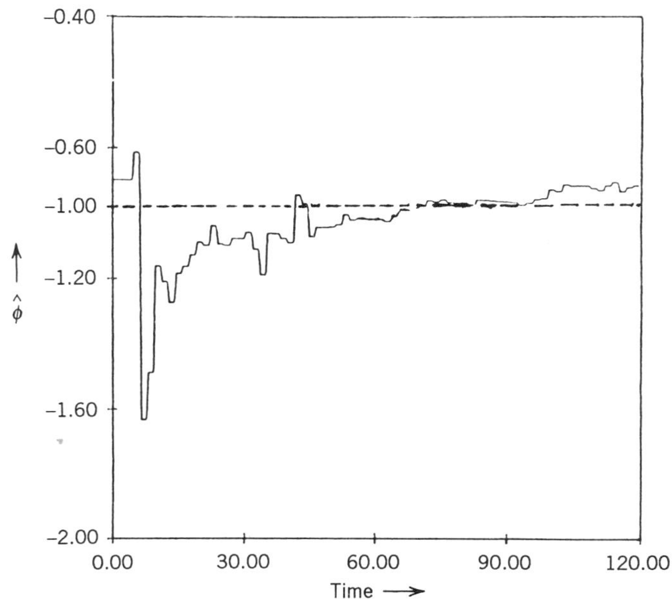

Fig. 6.2 Identification of parameter $\hat{\phi}$

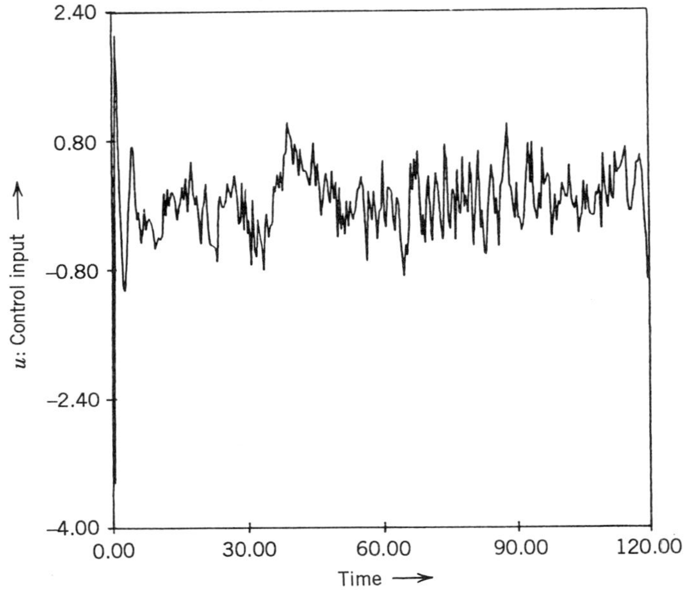

Fig. 6.3 Control input u

estimates of the Kalman filter equations:

$$
\begin{aligned}
d\hat{x}_1/dt &= \hat{x}_2 + p_{11}(z - \hat{x}_1), \\
d\hat{x}_2/dt &= \hat{\phi}\hat{x}_1 + p_{12}(z - \hat{x}_1), \\
dp_{11}/dt &= 2p_{12} - p_{11}^2, && p(0)_{11} = 10, \\
dp_{12}/dt &= p_{22} + \phi p_{12} - p_{11}p_{12}, && p(0)_{22} = 10, \\
dp_{22}/dt &= 2\phi p_{22} - p_{12}^2, && p(0)_{12} = 0,
\end{aligned}
$$

$$
H(\hat{x}) = \int_0^T \operatorname{Tr} P(t)\, dt = \int_0^T (p(t)_{11} + p(t)_{22})\, dt.
$$

Then the dual control cost is given by

$$
H(u^*) = H(u^*/\hat{x}) + H(\hat{x}).
$$

Now, for comparison purposes, the cost of identification $H(\hat{\phi})$ is removed from the cost of adaptive control. This cost is calculated from the results of Figure 6.2:

$$
H(\hat{\phi}) = \sum_{i=0}^{k} (\phi_i - \hat{\phi}_i)^2.
$$

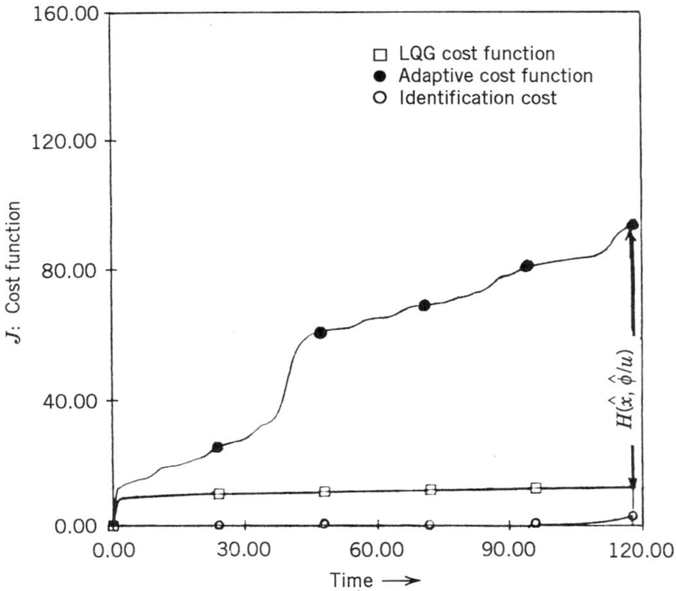

Fig. 6.4 LQG and adaptive cost functions

The resulting adaptive control cost is compared to the dual control cost in Figure 6.4:

$$H(u/\hat{\phi}) = H(u/\hat{x}, \hat{\phi}) + H(\hat{x}/\hat{\phi}) > H(u^*).$$

The difference between the two is the *equivocation of active transmission of information*, and as expected is positive, which proves that the adaptive control is not optimal.

This difference could serve as a measure of closeness of the adaptive to the dual control solution. The idea is further explored in the Section 6.6 of the present Chapter.

6.5 THE LQG OPTIMAL CONTROL AND THE KALMAN–BUCY FILTER

The difficulties arising from nonlinear optimal estimation and control are not present in the case of neutral systems, like linear-quadratic-Gaussian (LQG) systems given by

$$\begin{aligned} dx/dt &= F(t)x(t) + B(t)u(t) + G(t)w(t), \qquad x(0) \sim N(x_0, P_0), \\ z(t) &= H(t)x(t) + v(t), \qquad w(t) \sim N(0, Q);\, v(t) \sim N(0, R), \\ J(u) &= E\left\{\int_0^T [x^T M x + u^T N u]\, dt\right\} = H(u) = E\{I(u)\}. \end{aligned} \tag{6.51}$$

Selecting $p(x)$ to correspond to the minimum mean-square-error estimator Eq. (6.37), the Kalman–Bucy filter is derived from the associated worst-case differential entropy $H(\hat{x})$;

$$H(\hat{x}) = -\int_{\Omega_x} p(\hat{x}) \ln p(\hat{x})\, dx = \rho + \nu \int_{\Omega_x} \int_0^T \|\hat{x} - x\|^2\, dt\, p(\hat{x})\, dx$$
$$= \rho + \nu E\left\{ \int_0^T \|\hat{x} - x\|^2\, dt \right\}. \tag{6.52}$$

Minimization of the differential entropy implies

$$\min_x H(\hat{x}) \Leftrightarrow \min_x E\left\{ \int_0^T \left[\|\hat{x} - x\|^2 + \lambda^T F x + \lambda^T B u + \lambda^T G w - \lambda^T dx/dt \right] dt \right\} \tag{6.53}$$

This minimization problem is solved following Wonham's (1968) approach. Let

$$x(t) = x_1(t) + x_2(t),$$

such that

$$\begin{aligned} dx_1/dt &= F(t)x_1(t) + G(t)w(t), \qquad & x(0) &= x_1(0) + x_2(0) \\ dx_2/dt &= F(t)x_2(t) + B(t)u(t), \qquad & x_1(0) &\sim N(0, P_0) \\ & & x_2(0) &= x_0. \end{aligned} \tag{6.54}$$

If $\tilde{P}(t)$ is the error covariance matrix,

$$H(\hat{x}) = \rho + \nu E\left\{ \int_0^T \|\hat{x} - x\|^2\, dt \right\} = \rho + \nu E\left\{ \int_0^T \mathrm{Tr}[\hat{x} - x][\hat{x} - x]^T\, dt \right\}$$
$$= \rho + \nu E\left\{ \int_0^T \tilde{P}(t)\, dt \right\}. \tag{6.55}$$

Because x_2 is deterministic and known, $\hat{x}_2 = x_2$ and $\hat{x} = \hat{x}_1 + x_2$ is the total estimate. The measurement equations yield

$$\begin{aligned} z_1(t) &= z(t) - H(t)x_2(t) = H(t)x_1(t) + v(t), \\ z_2(t) &= x_2(t). \end{aligned} \tag{6.56}$$

The covariance is obtained as

$$E\{(\hat{x} - x)(\hat{x} - x)^T\} = E\{(\hat{x}_1 - x_1)(\hat{x}_1 - x_1)^T\} = \tilde{P}(t),$$

and the estimation problem is reduced to minimize $H(\hat{x}_1)$ subject to

$$dx_1/dt = F(t)x_1(t) + G(t)w(t),$$
$$z_1(t) = H(t)x_1(t) + v(t). \tag{6.57}$$

The optimal estimator is assumed to be of the following form:

$$d\hat{x}_1/dt = A_0(t)\hat{x}_1(t) + K(t)z_1(t), \qquad x(0) = 0,$$

the propagation matrix $A_0(t)$ is found to be

$$A_0(t) = F(t) - K(t)H(t), \tag{6.58}$$

and the filter gain $K(t)$ satisfies the matrix Riccati equation;

$$d\tilde{P}/dt = [F - KH]\tilde{P} + \tilde{P}[F - KH]^T + GQG^T + KRK^T, \qquad P(0) = P_0. \tag{6.59}$$

The value of $K(t)$ is obtained by minimizing Eq. (6.55) subject to Eq. (6.59) using variational methods, to yield

$$K(t) = \tilde{P}(t)H(t)^T R(t)^{-1}. \tag{6.60}$$

The estimator is then represented by

$$d\hat{x}/dt = \hat{x}_1(t) + x_2(t) = F(t)\hat{x}(t) + K(t)[z(t) - H(t)\hat{x}(t)],$$
$$\hat{x}(0) = x_0. \tag{6.61}$$

It was shown above that the optimal estimator is a Kalman–Bucy filter. It is claimed that the optimal controller now results from the minimization of $H(u/\hat{x})$. It may be interpreted as the optimal deterministic control minimizing $J(u)$ with the states replaced by their optimal estimates,

$$u^*(\hat{x}) = -A_1(t)\hat{x}(t), \tag{6.62}$$

where $A_1(t)$ is the optimal deterministic gain and the minimum value of the differential entropy is

$$H(u^*/\hat{x}) = \lambda + \mu J(u^*(\hat{x})). \tag{6.63}$$

To prove that $u^*(\hat{x})$ is really the optimal solution of the LQG it suffices to show that the equivocation $H(\hat{x}/u^*) = 0$ in the equation that follows, and therefore the system, is neutral, and the separation theorem holds [Saridis (1977)]:

$$H(u^*) = H(u^*/\hat{x}) + H(\hat{x}) - H(\hat{x}/u^*).$$

From the worst-case densities $p(\hat{x})$ and $p(u^*/\hat{x})$ the density $p(\hat{x}/u^*)$ is obtained, using the chain rule of probabilities:

$$p(\hat{x}/u^*) = \exp\left\{-\lambda - \rho - \mu[I(u^*(\hat{x})) - I(u(x))] - \nu\int_0^T \|\hat{x} - x\|^2\,dt\right\}, \tag{6.64}$$

and $\kappa = \lambda + \rho$; the equivocation is

$$\begin{aligned} H(\hat{x}/u^*) &= \kappa + \mu E\{[I(u^*(\hat{x})) - I(u(x))]\} - \nu E\left\{\int_0^T \|\hat{x} - x\|^2\,dt\right\} \\ &= \kappa - \mu\int_0^T \mathrm{Tr}[M(t)P(t)]\,dt + \nu\int_0^T \mathrm{Tr}\,P(t)\,dt. \end{aligned} \tag{6.65}$$

The last expression is deterministic constant and independent of $u^*(x)$ and x. Furthermore, the constants μ and v are arbitrary, so they may be selected to make the equivocation zero regardless of the selection of $u^*(\hat{x})$ and $\hat{x}$:

$$H(\hat{x}/u^*) \equiv 0. \tag{6.66}$$

This proves that the system is neutral, the separation theorem holds, and that $u^*(\hat{x})$ is truly the optimal control.

In the case of nonlinear systems, the quantity $E\{[I(u^*(\hat{x})) - I(u(x))]\}$ is *not* independent of $u^*(\hat{x})$, and therefore $H(\hat{x}/u^*)$ cannot be set to zero for all times $t \in [0, T]$ and constant μ and ν. This verifies Fel'dbaum's dual control.

6.6 UPPER BOUND OF THE EQUIVOCATION $H[\hat{\phi}/u^*]$

In the stochastic optimal adaptive control problem, when x is available for measurement, the equivocation is reduced to $H[\hat{\phi}/u^*]$. Because the calculation of $H[\phi/u^*]$ is practically impossible, it is easy to obtain an *upper bound*, assuming a least-squares identification algorithm, and invoking Jaynes's principle of maximum entropy, twice:

$$p(\hat{\phi}) = -\alpha - \beta\exp\left[-\|\hat{\phi} - \phi\|^2\right],$$

$$H\left[\hat{\phi}\right] = \alpha + \beta E\{\|\hat{\phi} - \phi\|^2\},$$

and

$$p(u/\hat{\phi}) = -\gamma - \sigma \exp[-V(x_0, t_0, u, \phi)],$$

$$H(u/\hat{\phi}) = \gamma + \sigma E\{V(x_0, t_0, u, \hat{\phi})\},$$

$$p(\hat{\phi}/u) = p(u/\hat{\phi})p(\hat{\phi})/p(u)$$

$$= \exp\Big\{-k - \sigma\, \partial V/\partial\phi(\hat{\phi} - \phi) + (\hat{\phi} - \phi)^T$$

$$\times\left[\tfrac{1}{2}\sigma\, \partial^2 V/\partial\phi^2 + I\right](\hat{\phi} - \phi) + o(\hat{\phi} - \phi)\Big\}, \quad (6.67)$$

to establish an *upper bound of the equivocation of active transmission of information*:

$$\overline{H}[\hat{\phi}/u] = C + E\left\{\sigma \frac{\partial V}{\partial \phi}(\hat{\phi} - \phi)\right.$$

$$\left. + (\hat{\phi} - \phi)^T \left[0.5\sigma \frac{\partial^2 V}{\partial \phi^2} + Q\right](\hat{\phi} - \phi)\right\} > H[\hat{\phi}/u]. \quad (6.68)$$

C denotes a normalization constant and $o(\cdot)$ is the little o notation. Recursive algorithms of the upper bound may be useful for on-line adaptive control algorithms.

Other self-organizing control algorithms described in Saridis (1977) may also be considered as entropy-minimization problems, and upper bounds of the equivocation may be useful in obtaining approximate solutions. They always provide a worst-case solution of the optimization problem.

EXAMPLE: THE UPPER BOUND OF EQUIVOCATION The following is an illustrative example of the use of approximate entropy measures to evaluate optimal and adaptive control solutions. For simplicity of calculations the following system, with only random initial conditions, is considered:

$$dx_1/dt = x_2, \qquad x(0) \sim N(0, 10I),$$

$$dx_2/dt = \phi x_2 + u, \qquad \phi = -1$$

$$z = x_1,$$

and the performance cost

$$J(u) = E\left\{\int_0^T (2x_1^2 + 2x_2^2 + \tfrac{1}{2}u^2)\, dt\right\}.$$

The problem is to compare the optimal and adaptive controls using approximate entropy measures, thus utilizing the upper bounds of equivocation to evaluate their performance for various initial conditions.

A. Optimal Control. Assuming that ϕ is known, and some fixed initial conditions x, the optimal solution is obtained by minimizing the entropy function:

$$\min_u H(x,u) = V(x) = p_{11}x_1^2 + 2p_{12}x_1x_2 + p_{22}x_2^2.$$

Using classical control solutions, the above yields

$$p_{11} = 3, \qquad p_{12} = 1, \qquad p_{22} = 1,$$

$$u^*(x) = -2x_1 - 2x_2,$$

and

$$V(x) = 3x_1^2 + 2x_1x_2 + x_2^2.$$

B. Adaptive Control. The adaptive control solution, assuming that ϕ is known, may be obtained by first identifying ϕ using a least-squares algorithm, and then substituting its estimated value to the deterministic optimal-control law. The least-squares identification algorithm generates ϕ, at discrete-time instants, sequentially;

$$\hat{\phi}(k) = \hat{\phi}(k-1) + P(k)z(k-1)\left[z(k) - z(k-1)\hat{\phi}(k-1)\right],$$

$$P(k) = P(k-1) - P^2(k-1)z^2(k-1)/\left[P^2(k-1)z^2(k-1) + 1\right],$$

which minimizes

$$I = E\left\{\left\| z(k) - z(k-1)\hat{\phi}\right\|^2\right\}.$$

The optimal-control law is obtained by minimizing the deterministic Hamilton–Jacobi–Bellman equation (6.18), using the estimated parameter ϕ:

$$u^*(x) = -2x_1 - \left[\hat{\phi} + \sqrt{\hat{\phi}^2 + 8}\right]x_2.$$

The value of $V(x,\hat{\phi}) = V(x,u^*,\hat{\phi})$ for $u^*(x)$ and $\hat{\phi}$ is given by

$$V(x,\hat{\phi}) = \left[9 + \hat{\phi} + \hat{\phi}^2 + (\hat{\phi}+1)\sqrt{\hat{\phi}^2+8}\right]x_1^2\Big/\left[\hat{\phi} + 1 + \sqrt{\hat{\phi}^2+8}\right] + 2x_1x_2$$

$$+\left[4 + \tfrac{1}{2}\hat{\phi}^2 + \tfrac{1}{2}\hat{\phi}\sqrt{\hat{\phi}^2+8}\right]x_2^2\Big/\left[\hat{\phi} + 1 + \sqrt{\hat{\phi}^2+8}\right].$$

The values of the performance cost $V(x, \hat{\phi})$ are sequentially upgraded as new estimates of ϕ are obtained.

C. Upper Bound of the Equivocation. The equivocation of active transmission of information has been defined to be a function of the probability density:

$$p(\hat{\phi}/u) = \exp\left\{-\sigma\left[V(x,\hat{\phi}) - V(x,\phi)\right] - \|\hat{\phi} - \phi\|^2\right\}$$

and

$$H(\hat{\phi}/u) = E\left\{\sigma\left[V(x,\hat{\phi}) - V(x,\phi)\right] + \|\hat{\phi} - \phi\|^2\right\}.$$

It is practically impossible to calculate this function. Instead, the upper bound of the equivocation is calculated and used as a measure to compare the optimal and adaptive solutions:

$$H_{\text{ub}} = E\left\{|\sigma\,\partial V/\partial\phi \cdot D\phi| + \tfrac{1}{2}\left[|\sigma\,\partial^2 V/\partial\phi^2| + |I|\right] D\phi^2\right\},$$

where the difference between estimate and true value $D\phi$ is defined as

$$D\phi = \hat{\phi} - \phi.$$

For $\sigma = 1$, the components of the upper bound [Eq. (6.68)] are

$$\partial V/\partial\phi = \frac{\left[2\hat{\phi}^3 + 3\hat{\phi}^2 + 8 + 2\hat{\phi}(\hat{\phi}+1)\sqrt{\hat{\phi}^2+8}\right]}{\left[2\hat{\phi}^3 + 2\hat{\phi}^2 + 16\hat{\phi} + \left(2\hat{\phi}^2 + 2\hat{\phi} + 9\right)\sqrt{\hat{\phi}^2+8}\right]\left[x_1 + \tfrac{1}{2}x_2^2\right]}$$

and

$$\partial^2 V/\partial\phi^2 = \frac{\begin{aligned}&\Big[-2\hat{\phi}^4 + 86\hat{\phi}^3 + 190\hat{\phi}^2 + 206\hat{\phi}\\ &\quad + \left(52\hat{\phi}^5 + 116\hat{\phi}^4 + 33\hat{\phi}^3 + 32\hat{\phi}^2 + 36\hat{\phi} + 96\right)\sqrt{\hat{\phi}^2+8}\Big]\end{aligned}}{\begin{aligned}&\Big[8\hat{\phi}^6 + 16\hat{\phi}^5 + 140\hat{\phi}^4 + 228\hat{\phi}^3 + 721\hat{\phi}^2 + 800\hat{\phi} + 904\\ &+ \left(8\hat{\phi}^5 + 16\hat{\phi}^4 + 108\hat{\phi}^3 + 164\hat{\phi}^2 + 352\hat{\phi} + 128\right)\sqrt{\hat{\phi}^2+8}\Big]\left[x_1^2 + \tfrac{1}{2}x_2^2\right]\end{aligned}}$$

The upper bound is then evaluated and compared to the actual equivocation for values of the parameter ϕ, and different initial conditions x_{10}, x_{20}, in Table 6.1.

In this table 1C stands for initial conditions x_0, $X^2 = (x_{10}^2 + x_{20}^2)$, and

$$DV = V_{\text{ad}} - V_{\text{opt}}.$$

TABLE 6.1 Upper-Bound Evaluation for Various Initial Conditions

Initial Condition	Optimal	Adaptive	DV	$\lvert\partial V/\partial\phi\rvert$	$\lvert\partial^2 V/\partial\phi^2\rvert$	X^2	$\overline{H}$
$\phi = -0.50$, $D\phi = 0.5$							
IC 10	600.0	610.25	10.25	0.216	0.204	150.0	20.6
IC 5	150.0	152.50	2.50	0.216	0.204	37.5	5.325
IC 1	6.0	6.30	0.30	0.216	0.204	1.5	0.453
$\phi = -0.75$, $D\phi - 0.25$							
IC 10	600.0	607.2	7.2	0.27	0.22	150.0	11.22
IC 5	150.0	151.95	1.95	0.27	0.22	37.5	2.85
IC 1	6.0	6.27	0.27	0.27	0.22	1.5	0.33
$\phi = -0.9$, $D\phi = 0.1$							
IC 10	600.0	602.10	2.20	0.33	0.232	150.0	5.13
IC 5	150.0	150.60	0.60	0.33	0.232	37.5	1.29
IC 1	6.0	6.04	0.04	0.33	0.232	1.5	0.06

This example demonstrates the ability of obtaining a measure of closeness between the adaptive and the optimal solutions. The acceptance of the adaptive solution as approximately optimal during the transient period depends on the acceptance of the calculated difference DV.

6.7 CASE STUDY ON OPTIMAL CONTROL OF INTELLIGENT SPACE TRUSS ASSEMBLY

The application of the entropy theory to stochastic control problems will be demonstrated by a case study that uses entropy to define a uniform measure of performance of various levels of an Intelligent Control system, developed for optimal control of a space truss assembly.

The theory of Intelligent Control systems, developed by Saridis, combines the powerful high-level decision making of the digital computer with advanced mathematical modeling and synthesis techniques of system theory with linguistic methods of dealing with imprecise or incomplete information [Saridis and Valavanis (1988)]. This produces a unified approach for the design of Intelligent Machines. The theory may be thought of as the result of the intersection of the three major disciplines of Artificial Intelligence, Operations Research, and Control Theory. The theory is aimed at establishing Intelligent Controls as an engineering discipline, with the purpose of designing Intelligent Autonomous Systems of the future.

Intelligent Control provides the fusion between the mathematical and linguistic methods and algorithms applied to systems and processes. It effectively combines the results of cognitive systems research, with various mathematical programming control techniques.

The control intelligence is hierarchically distributed according to the *Principle of Precision with Decreasing Intelligence (IPDI)*, evident in all hierar-

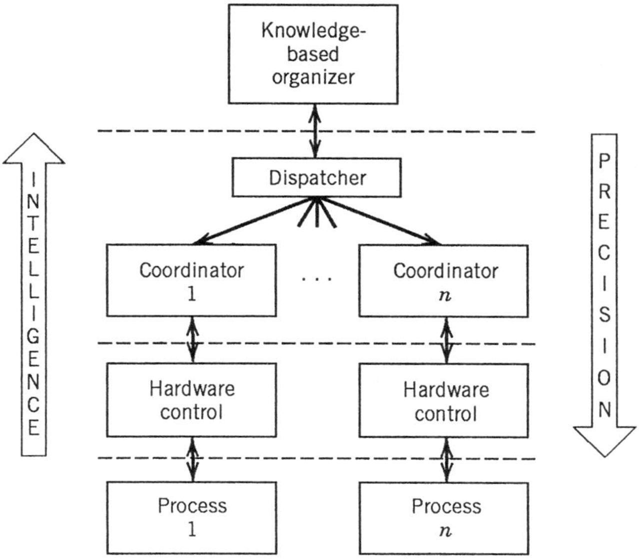

Fig. 6.5 A hierarchically intelligent control system

chical management systems [Saridis (1988)]. The analytic functions of an intelligent machine are implemented by intelligent controls, using entropy as a measure. The resulting structure is composed of three basic levels of controls, each level of which may contain more than one layer of tree-structured functions, as depicted in Figure 6.5:

The *organization level* is modeled after a Boltzmann machine for abstract reasoning, task planning, and decision making.

The *coordination level* is composed of a number of Petri Net Transducers supervised, for command exchange, by a dispatcher, which also serves as an interface to the organization level.

The *execution level* includes the sensory, planning for navigation, and control hardware that interacts one-to-one with the appropriate coordinators, and a VME bus provides a channel for data-base exchange among the several devices.

The functions involved in the upper levels of an intelligent machine are imitating functions of human behavior and may be treated as elements of knowledge-based systems. Actually, the activities of planning, decision making, learning, data storage and retrieval, task coordination, et cetera may be thought of as knowledge handling and management [Valavanis and Saridis (1992)].

Such a system was implemented to perform autonomously construction of NASA's Space Station Freedom at the Center of Intelligent Robotic Systems

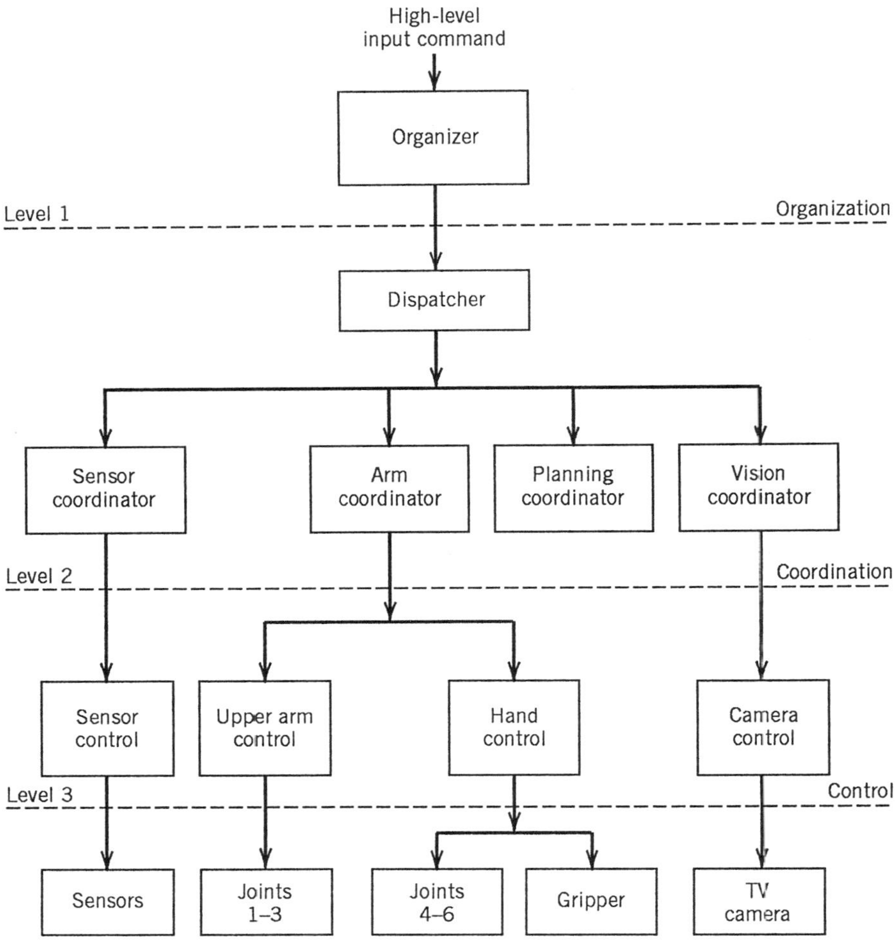

Fig. 6.6 A hierarchically intelligent control for a manipulator with vision and sensory feedback

for Space Exploration (CIRSSE) at RPI. The entropy measures of performance for the optimization of the last two levels of the system shown in Figure 6.6, are presented in the following.

6.7.1 Architecture of the Coordination Level

The *Coordination level* is a tree structure of *Petri Net Transducers* as coordinators, proposed by Wang and Saridis (1990) with the Dispatcher as the root. It is shown in Figure 6.6. The Petri Net Transducer for the Dispatcher is generated by the Organizer for every specific plan and is transmitted, asynchronously, to the Coordination level along with the plan to be executed. The function of the Dispatcher is to interpret the plan and assign individual tasks to the other coordinators, monitor their operation, and transmit messages and commands from one coordinator to another as needed. As an example, a command is sent to the vision and sensing coordinator to generate a model of the environment and determine the coordinates of the objects for manipulation to be tabulated, and then transmitted to the motion coordinator for navigation and motion control. This command is executed by having each transition of the associated Petri nets to initialize a package corresponding to a specific action. These packages are stored in short memories associated with each of the coordinators.

The rest of the coordinators have a fixed structure with alternate menus available at request. They communicate commands and messages with each other, through the Dispatcher. They also provide information about reception of a message, data memory location, and job completion. No data are communicated at the Coordination level, because the task planning and monitoring may be located in a remote station, and such an exchange may cause a channel congestion. A preferred configuration for such situations is that the coordinators with a local dispatcher may be located with the hardware at the work site, while a remote dispatcher, connected to the organizer, interacts with local one from a remote position. This concept simplifies the communication problem considerably, because only short messages are transmitted back and forth through a major channel between local and remote stations, requiring a narrow bandwidth. An example of the effectiveness of such an architecture may be demonstrated in space construction, where robots work in space while task planning and monitoring is done on Earth.

Even though there is no limitation to the number of coordinators attached to the Dispatcher, only the following ones are planned for an Intelligent Robot for space truss assembly.

Vision Coordinator. This device coordinates all the sensory activities of the robot, with cameras and lasers, and generates information of the world model in Cartesian coordinates.

Sensory Coordinator. This device coordinates mainly the activities of proximity sensors, for fine motion control.

Motion Control Coordinator. This device receives control, object, and obstacle information and uses it to navigate and move multiple robotic arms and other devices, for object manipulation and task execution. It also assigns the appropriate operations on the data acquired for the desired application.

Entropy measures are developed by McInroy and Saridis (1991) at each coordinator such that they may be used to minimize the complexity and improve the reliability of the system.

6.7.2 Analytic Model

Petri nets have been proposed as devices to communicate and control complex heterogenous processes. These nets provide a communication protocol among stations of the process as well as the control sequence for each one of them. Abstract task plans suitable for many environments are generated at the organization level by a grammar created by Wang and Saridis (1990):

$$G = (N, \Sigma_0, P, S) \tag{6.69}$$

where

$$N = \{S, M, Q, H\} = \text{nonterminal symbols}$$
$$\Sigma_0 = \{A_1, A_2 \cdots A_n\} = \text{terminal symbols (activities)}$$
$$P = \text{production rules.}$$

Petri net transducers (PNTs) proposed first by Wang and Saridis, are Petri net realizations of the *Linguistic Decision Schemata* introduced by Saridis and Graham (1984), as linguistic decision-making and sequencing devices. They are defined as 6-tuples:

$$M = (N, \Sigma, \delta, G, \mu, F), \tag{6.70}$$

where

$$N = (P, T, I, O) = \text{a Petri net with initial marking } \mu,$$
$$\Sigma = \text{a finite input alphabet}$$
$$\delta = \text{a finite output alphabet}$$
$$\sigma = \text{a translation mapping from } T \times (\Sigma \cup \{\backslash\}) \text{ to finite sets of } \delta^*$$
$$\text{and } F \subset R(\mu) \text{ a set of final markings.}$$

A *Petri net transducer* (PNT) is a Petri net with input and output tapes. It communicates with the other levels through input and output languages called *Petri Net Languages* (PNL). In addition to its on-line decision-making capability PNTs have the potential of generating communication protocols and learning by feedback, ideal for the communication and control of coordinators and their dispatcher in real time. Their architecture is given in Figure 6.7, and may follow a scenario suitable for the implementation of an autonomous intelligent robot.

Decision making in the coordination structure is accomplished by *Task Scheduling* and *Task Translation* [Wang and Saridis (1990)]. The sequence of events transmitted from the organization level is received by the dispatcher, which requests a world model with coordinates from a vision coordinator.

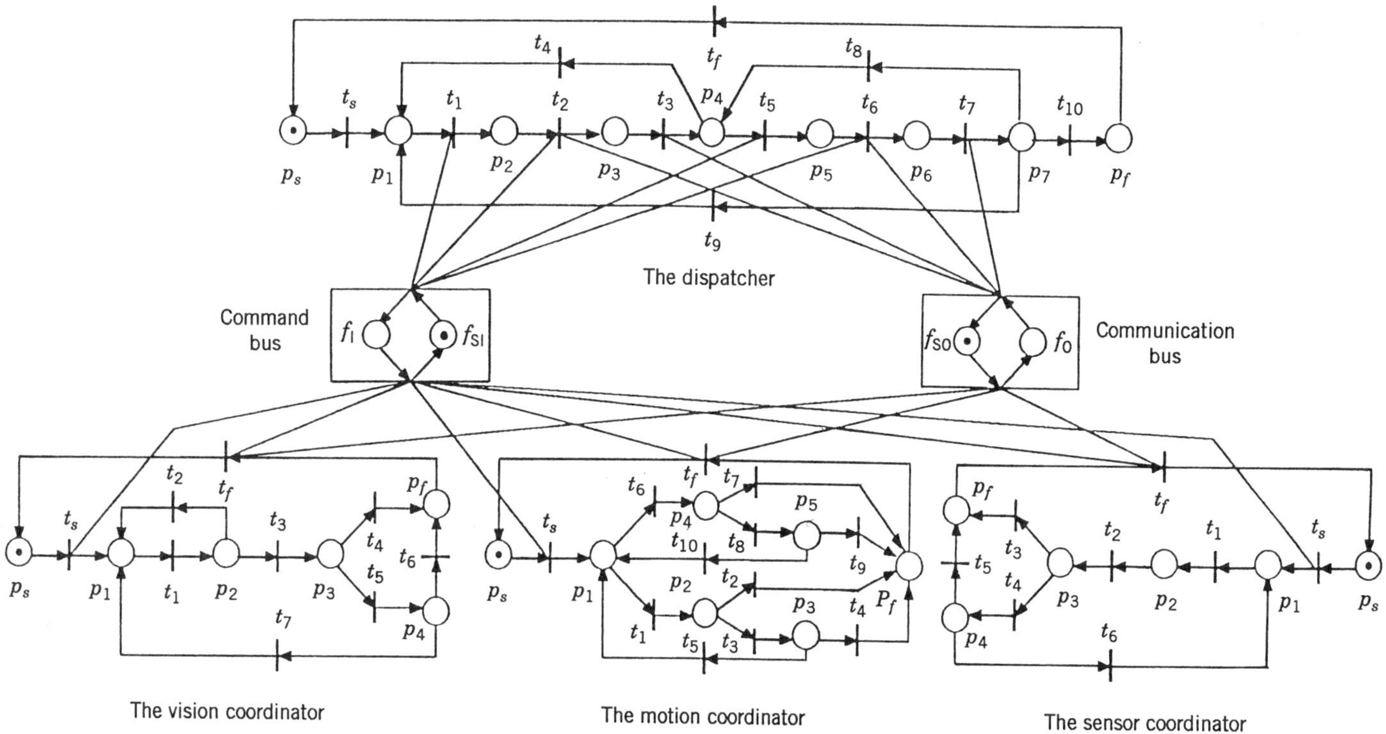

Fig. 6.7 The coordination structure of intelligent manipulator system

The vision coordinator generates an appropriate data-base, and upon the dispatcher's command communicates it to the planning coordinator, which sets a path for the arm manipulator. A new command from the dispatcher sends path information to the motion controller in terms of end points, constraint surface, and performance criteria. The vision coordinator is then switched to a monitoring mode for navigation control, and so on.

The PNTs can be evaluated in *real-time* by testing the computational complexity of their operation, which may be expressed uniformly in terms of entropy.

Feedback information is communicated to the coordination level from the execution level during the execution of the applied command. Each coordinator, when accessed, issues a number of commands to its associated execution devices (at the execution level). Upon completion of the issued commands feedback information is received by the coordinator and is stored in the *short-term memory* of the coordination level. This information is used by other coordinators if necessary, and also to calculate the individual, accrued, and overall accrued costs related to the coordination level. Therefore, the feedback information from the execution to the coordination level will be called *on-line real-time* feedback information.

The performance estimate and the associated subjective probabilities are updated after the k_{ij}th execution of a task $[(u_t, x_t)_i, S_j]$ and the measurement of the estimate of the observed cost J_{ij}:

$$\begin{aligned} J_{ij}(k_{ij}+1) &= J_{ij}(k_{ij}) + \beta(k_{ij}+1)\left[J_{\text{obs}}(k_{ij}+1) - J_{ij}(k_{ij})\right], \\ P_{ij}(k_{ij}+1) &= P_{ij}(k_{ij}) + \mu(k_{ij}+1)\left[\Gamma_{ij}(k_{ij}+1) - P_{ij}(k_{ij})\right], \end{aligned} \tag{6.71}$$

where

$$\Gamma_{ij} = \begin{cases} 1, & \text{if } J_{ij} = \min, \\ 0, & \text{elsewhere}, \end{cases}$$

and β and μ are harmonic sequences. Convergence of this algorithm is proven in Saridis and Graham (1984).

The *learning process* is measured by the entropy associated to the subjective probabilities:

$$H(\mathrm{C}) = H(\mathrm{E}) + H(\mathrm{T}/\mathrm{E}), \tag{6.72}$$

where H(E) is the environmental uncertainty and H(T/E) is the pure translation uncertainty. Only the last term can be reduced by learning.

6.7.3 Architecture of the Execution Level

The *execution level* contains all the hardware required by the intelligent machine to execute a task. There is a one-to-one correspondence between hardware groups and coordinators. Therefore their structure is usually fixed.

This level also contains all the drivers, VME buses, short memory units, processors, actuators, and special-purpose devices needed for the execution of a task. After the successful completion of a job, feedback information is generated at this level for evaluation and parameter updating of the whole machine. Complexity dominates the performance of this level. Because *precision* is proportional to *complexity*, it also defines the amount of effort required to execute a task. It has been shown that all the activities of this level can be measured by entropy, which may serve as a measure of complexity as well. Minimization of local complexity through feedback may serve as a local design procedure.

The localization of data exchange at this level provides a means of efficient remote control of the Intelligent Machine. Because of the diversity of the hardware in a general purpose intelligent machine, this work will focus on the special case of a robot designed for space construction like the CIRSSE transporter. The following hardware groups are available.

The Vision System. This system consists of two cameras fixed at the ceiling of the laboratory, two penlight cameras on the wrist of one PUMA arm, and a laser range finder. They are all controlled by a Datacube with a versatile menu of various hard-wired functions and a VME bus for internal communications. The functions assigned to them, for example, create a world model in Cartesian space, find the fiducial marks on the object to be manipulated, or track a moving object, are supported by software specialized for the hardware of the system. Calibration and control of the hardware is an important part of the system. Because we are dealing with information processing, the system's performance can be easily measured with entropy. Actual data for visual servoing can be generated on the VME bus and transmitted through the Dispatcher to the Motion Control system. Internal communication on a VME bus and direct connection with the VME bus of the Motion Control system has been developed as the Vision Servicing System (VSS) system.

The Sensory System. This system consists of two proximity sensors that are used for control of fine motion of the manipulator and the gripper. It functions in a manner very similar to the vision system, but it acts independent of the motion control and vision system buses.

The Motion Control System. This system is a unified structure for cooperative motion and force control for multiple arm manipulation. Because motion affects force but not vice versa, motion control is designed independent of the constraint forces, and force control by treating inertial forces as disturbance. Integral force feedback is used with full dynamics control algorithms. The resulting system, named CTOS for CIRSSE Testbed Operating System, was developed as a multiple-processor, VME-bus-based, real-time robot control system for the CIRSSE 18-degree-of-freedom transporter. It hierarchically integrates the execution algorithms in planning, interaction, and servo control. It works together with the VXWORKS software and provides most of the transformations, and other kinematics and dynamics tools needed for servoing and manipulation. In earlier work it was shown that the control activities can be measured by entropy [Saridis (1988)]. Therefore the measure of performance of the Motion Control System is consistent with the rest of the architecture of the Intelligent Machine.

6.7.4 Entropy Formulation of Motion Control

The cost of control at the hardware level can be expressed as an entropy that measures the uncertainty of selecting an appropriate control to execute a task. By selecting an optimal control, one minimizes the entropy, for example, the uncertainty of execution. The entropy may be viewed in this respect as an energy in the original sense of Boltzmann (1872). Optimal-control theory utilizes a nonnegative functional of the state of the system $x(t) \in \Omega_x$, the state space, and a specific control $u(x,t) \in \Omega_u \times T$; $\Omega_u \subset \Omega_x$ the set of all admissible feedback controls, to define the performance measure for some initial conditions $x_0(t_0)$, representing a generalized energy function, of the form

$$V(x_0, t_0) = E\left\{\int_{t_0}^{t_f} L(x, t; u(x,t))\, dt\right\}, \tag{6.73}$$

where $L(x, t; u(x,t)) > 0$, subject to the differential constraints dictated by the underlying process

$$dx/dt = f(x, u(x,t), w, t), \qquad x(t_0) = x_0$$
$$z = g(x, v, t), \quad x(t_f) \in M_f, \tag{6.74}$$

where x_0, $w(t)$, and $v(t)$ are random variables with associated probability densities $p(x_0)$, $p(w(t))$, $p(v(t))$, and M_f a manifold in Ω_x. The trajectories of the system (6.70) are defined for a fixed but arbitrarily selected control $u(x,t)$ from the set of admissible feedback controls Ω_u.

In order to express the control problem in terms of an entropy function, one may assume that the performance measure $V(x_0, t_0, u(x,t))$ is distributed in u according to the probability density $p(u(x,t))$ of the admissible controls $u(x,t) \in \Omega_u$. The *differential entropy* $H(u)$ corresponding to the density is defined as

$$H(u) = -\int_{\Omega_u} p(u(x,t)) \ln p(u(x,t))\, dx,$$

and represents the uncertainty of selecting a control $u(x,t)$ from all possible admissible feedback controls Ω_u. The optimal performance should correspond to the maximum value of the associated density $p(u(x,t))$. Equivalently, the optimal control $u^*(x,t)$ should minimize the entropy function $H(u)$. This is satisfied if the density function is selected to satisfy *Jaynes' principle of maximum entropy* (1957), for example,

$$p(u(x,t)) = \exp\{-\lambda - \mu V(x_0, t_0; u(x,t))\}, \tag{6.75}$$

where λ and μ are normalizing constants.

It was shown by Saridis (1988) that the expression $H(u)$ representing the entropy for a particular control action $u(x,t)$ is given by

$$\begin{aligned} H(u) &= \int_{\Omega_u} p(x,t;u(x,t))V(x_0,t_0;u(x,t))\,dx \\ &= \lambda + \mu V(x_0,t_0;u(x,t)). \end{aligned} \tag{6.76}$$

This implies that the average performance measure of a feedback control problem corresponding to a specifically selected control is an entropy function. The optimal control $u^*(x,t)$ that minimizes $V(x_0,t_0;u(x,t))$, maximizes $p(x,t;u(x,t))$, and consequently minimizes the entropy $H(u)$:

$$u^*(x,t) \colon E\{V(x_0,t_0;u^*(x,t))\} = \min \int_{\Omega_u} V(x_0,t_0;u(x,t))p(u(x,t))\,dx. \tag{6.77}$$

This statement is the generalization of a theorem proven by Saridis (1988), and establishes equivalent measures between information theoretic and optimal-control problems and provides the information and feedback control theories with a common measure of performance.

6.7.5 Entropy Measure of the Vision System

The optimal-control theory designed mainly for motion control can be implemented for vision control, path planning, and other sensory systems pertinent to an Intelligent Machine by slightly modifying the system equations and cost functions. After all, one is dealing with real-time dynamic systems that may be modeled by a dynamic set of equations.

A *Stereo Vision* system of a pair of cameras mounted at the end of a robot arm may be positioned at $i = 1 \cdots N$ different view points to reduce problems with noise, considered one at a time due to time limitations. The accuracy of measuring the object's position depends upon its relative position in the camera frame. Consequently, each viewpoint will have different measurement error and time statistics. These statistics may be generated to define the uncertainty of the measurement of the vision system as in McInroy and Saridis (1991).

For a point c of the object, the measurement error of its 3D position in the camera coordinate frame e_{pc} is given by

$$e_{pc} = M_c n_c \tag{6.78}$$

where n_c is the 3D image position errors, and M_c an appropriate 3×3 matrix, depending on the position of the object. The linearized orientation error is given by

$$\delta = (M^T M)^{-1} M^T M' F n, \tag{6.79}$$

where

- δ is the orientation error in the camera frame,
- M is a matrix formed from camera coordinate frame positions,
- M' is a constant matrix,
- F is the matrix formed from the camera parameters and measured positions,
- n is the vector of the image position errors at the four points.

A vector containing the position and orientation errors due to image noise is given by

$$e_c = \left[e_{pc}^T\ \delta^T\right]^T = Ln, \tag{6.80}$$

where L depends on the camera parameters and the four measured camera frame positions of the points. The statistics of the image noise n due to individual pixel errors are assumed to be uniformly distributed. Assuming that feature-matching centroids are used by the vision system, its distributions tend to be independent Gaussian, due to the Central Limit theorem:

$$n \approx N(0, C_v) \quad \text{and} \quad e_c \approx N\left(0, LC_v L^T\right). \tag{6.81}$$

The time each vision algorithm consumes is also random, due to the matching period. Therefore the total vision time, for the ith algorithm, which includes camera-positioning time, image-processing time, and transformation to the base frame, is assumed Gaussian:

$$t_{vi} \approx N\left(\mu_{tvi}, \sigma_{tvi}^2\right). \tag{6.82}$$

Once the probability density functions are obtained, the resulting entropies $H(t_{vi})$, and $H(e_c)$, are obtained in a straightforward manner for the ith algorithm [McInroy and Saridis (1991)]:

$$H(t_{vi}) = \ln \sqrt{(2\pi e \sigma_{tvi}^2)},$$

$$H(e_c) = \ln \sqrt{(2\pi e)^6 \det[C_v]} + E\{\ln[\det L_i]\}. \tag{6.83}$$

The total entropy may be used as a measure of uncertainty of the Vision system (imprecision), and can be minimized with respect to the available system parameters:

$$H(V) = H(t_{vi}) + H(e_c). \tag{6.84}$$

6.7.6 Entropy Measure for the Sensory System

The proximity sensory system has similar properties as the vision system. Therefore its entropy measure should be similar to Eq. (6.84):

$$H(\mathrm{S}) = H(t_{vi}) + H(e_c). \tag{6.85}$$

6.7.7 Total Entropy of the System

Because the Coordination level works on a different time scale than the Execution level, the total entropy of the system is obviously the sum of the

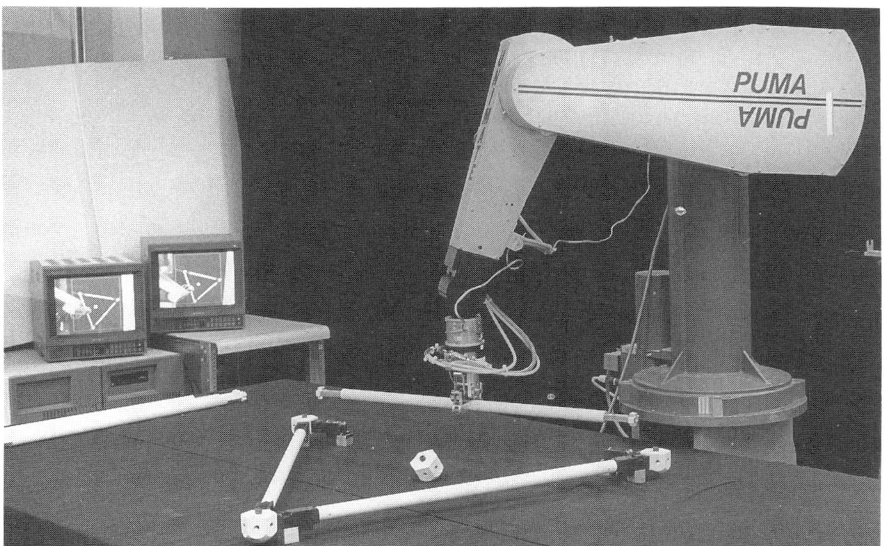

Fig. 6.8 Truss assembly using hierarchically intelligent control

partial entropies of the Coordination $H(\mathrm{C})$, the Motion Control $H(\mathrm{MC})$, the Vision system $H(\mathrm{V})$, and the Sensory system $H(\mathrm{S})$:

$$H(\mathrm{TOTAL}) = H(\mathrm{C}) + H(\mathrm{MC}) + H(\mathrm{V}) + H(\mathrm{S}). \qquad (6.86)$$

By selecting individual controls to minimize the total entropy $H(\mathrm{TOTAL})$ one obtains the optimal truss assembly procedure for space construction.

Figure 6.8 is a photograph of the experimental system for truss assembly designed and operated at the Center for Intelligent Robotic Systems for Space Exploration at the Rensselaer Polytechnic Institute in 1993.

6.8 CONCLUSIONS

The entropy formulations of optimal control, state estimation, and parameter identification were presented in this chapter for continuous-time systems only. The discrete-time version of the same problem is the same as the one presented and was omitted to avoid repetition. It will be given, though, as a set of homework problems in the following.

The entropy formulation of optimal control, state estimation, and parameter identification presents a new approach to solve these problems, and at the same time reformulates all the solutions from a global point of view, providing a common measure with other heterogeneous disciplines. In addition, it provides an analytic proof that the active feedback (Dual) control is the globally optimal control, and provides a measure for evaluation of its difference with the Open-Loop-Feedback-Optimal and Adaptive Control solutions. This proves to be a very powerful tool in the theory of optimal control.

A space assembly project, developed at CIRSSE of RPI, and used as a case study, demonstrated the usefulness of using entropy as a unifying measure, for optimal control of a process, combining various disjoint disciplines.

PROBLEMS

6.1 Derive the Kalman–Bucy filter using entropy concepts for the continuous-time system;

$$\begin{aligned} dx_1/dt &= x_2(t), & x(0) &\sim N(1, 10I), \\ dx_2/dt &= -x_2(t) + u(t) + w(t), & w(t) &\sim N(0,1), \\ z(t) &= x_1(t) + v(t), & v(t) &\sim N(0,1), \end{aligned}$$

and $u(t) = 1$.

6.2 For the system in Problem 6.1, find the optimal control for a performance cost

$$J(u) = E\left\{\int_0^T [x_1^2 + x_2^2]\, dt.\right.$$

6.3 Derive the Kalman filter using entropy concepts for the discrete-time system:

$$x(k+1) = Fx(k) + Gw(k), \qquad x(0) \sim N(x_0, P_0),$$

$$z(k) = Hx(k) + v(k), \qquad w(k) \sim N(0, Q), v(k) \sim (0, R).$$

6.4 Find the optimal control for the discrete-time system;

$$x(k+1) = Fx(k) + Bu(k) + Gw(k), \qquad x(0) \sim N(x_0), P_0)$$

$$z(k) = Hx(k) + v(k), \qquad w(k) \sim N(0, Q), \qquad v(k) \sim (0, R)$$

and performance cost

$$J(u) = E\left\{\sum_0^{n-1} \left[x(i+1)^T Mx(k+1) + u(i)^T Nu(i)\right]\right\}.$$

REFERENCES

Ashby, W. R., *An Introduction to Cybernetics*, Wiley, New York, 1975.

Boettcher, K. L., and Levis, A. H., "Modeling the Interacting Decision-Maker with Bounded Rationality," *IEEE Trans. Systems Man Cybern.*, Vol. SMC-12, No. 3, 1983.

Boltzmann, L., "Further Studies on Thermal Equilibrium Between Gas Molecules," *Wien Ber.*, Vol. 66, p. 275 (1872).

Conant, R. C., "Laws of Information which Govern Systems," *IEEE Trans. Systems Man Cybern.*, Vol. SMC-6, No. 4, pp. 240–255 (1976).

Feld'baum, A. A., *Optimal Control Systems*, Academic Press, New York, 1965.

Jaynes, E. T., "Information Theory and Statistical Mechanics," *Phys. Rev.*, Vol. 4, p. 106 (1957).

Kalata, P., and Premier, R., "On Minimal Error Entropy Stochastic Approximation," *Int. J. Syst. Sci.*, Vol. 5, No. 9, pp. 985–986 (1974).

Kolmogorov, A. N., "On Some Asymptotic Characteristics of Completely Bounded Metric Systems," *Dokl. Akad. Nauk., SSSR*, Vol. 108, No. 3, pp. 385–389 (1956).

Kumar, P. R., and Varaiya, P., *Stochastic Systems: Estimation, Identification, and Adaptive Control*, Prentice Hall, Englewood Cliffs, NJ, 1986.

Lindsay, R. B., and Margenau H., *Foundations of Physics*, Dover Publications, New York, NY, 1957.

McInroy, J. E., and Saridis, G. N., "Reliability Based Control and Sensing Design for Intelligent Machines," in J. H. Graham (Ed.), *Reliability Analysis*, Elsevier North Holland, New York, 1991.

Morse, A. S., "Towards a Unified Theory of Parameter Adaptive Control: Tunability," *IEEE Trans. Autom. Control*, Vol. AC-35, No. 9, pp. 1002–1012 (1990).

Morse, A. S., "Towards a Unified Theory of Parameter Adaptive Control—Part II: Certainty Equivalence and Implicit Tuning," *IEEE Trans. Autom. Control*, Vol. AC-37, No. 1, pp. 15–29 (1992).

Prigogine, I., *From Being to Becoming*, W. H. Freeman and Co., San Francisco, CA, 1980.

Saridis, G. N., *Self-Organizing Controls of Stochastic Systems*, Marcel Dekker, New York, 1977.

Saridis, G. N., "Toward the Realization of Intelligent Controls," *IEEE Proc.*, Vol. 67, No. 8 (1979).

Saridis, G. N., "Intelligent Robotic Control," *IEEE Trans. on Autom. Control*, Vol. AC-28, No. 4, pp. 547–557 (1983).

Saridis, G. N., "Intelligent Control: Operating Systems in Uncertain Environments," in J. Ackermann (Ed.), *Uncertainty and Control*, Springer Verlag, Berlin, 1985a, Chap. 7, pp. 215–233.

Saridis, G. N., "An Integrated Theory of Intelligent Machines by Expressing the Control Performance as an Entropy," *Control Theory Adv. Technol.*, Vol. 1, No. 2, pp. 125–138 (1985b).

Saridis, G. N., "Foundations of Intelligent Controls," in *Proceedings of IEEE Workshop on Intelligent Controls*, RPI, Troy, NY, 1985c.

Saridis, G. N., "Entropy Formulation for Optimal and Adaptive Control," *IEEE Trans. Autom. Control*, Vol. AC-33, No. 8, pp. 713–721 (1988).

Saridis, G. N., "Analytic Formulation of the IPDI for Intelligent Machines," *Automatica* Vol. 25, No. 3, pp. 461–467 (1989).

Saridis, G. N., "Architectures for Intelligent Machines," in *Proceedings of Workshop for Advanced Robotics*, Beijing, People's Republic of China, August 23–29, 1991.

Saridis, G. N., and Graham, J. H., "Linguistic Decision Schemata for Intelligent Robots," *Automatica*, Vol. 20, No. 1, pp. 121–126 (1984).

Saridis, G. N., and Lee, C. S. G., "Approximation of Optimal Control for Trainable Manipulators," *IEEE Trans. Systems Man Cybern.*, Vol. 8, No. 3, pp. 152–159 (1979).

Saridis, G. N., and Valavanis, K. P., "Analytical Design of Intelligent Machines," *Automatica* Vol. 24, No. 2, pp. 123–133 (1988).

Shannon, C., and Weaver, W., *The Mathematical Theory of Communications*, Illini Books, 1963.

Tsai, Y. A., Casiello, F. A., and Loparo, K. A., "Discrete-Time Entropy Formulation of Optimal and Adaptive Control Problems," *IEEE Trans. Autom. Control*, Vol. AC-37, No. 7, pp. 1083–1088 (1992).

Valavanis, K. P., Saridis, G. N., *Intelligent Robotic System Theory: Design and Applications*, Kluwer Academic Publishers, Boston, 1992.

Wang, F., and Saridis, G. N., "A Coordination Theory for Intelligent Machines," *Automatica* Vol. 35, No. 5, pp. 833–844 (1990).

Weidemann, H. L., "Entropy Analysis of Feedback Control Systems," in *Advances in Control Systems*, C. Leondes, Ed., Academic Press, New York, NY, 1969.

Wonham, W. M. "On the Separation Theorem of Stochastic Control" SIAM Journal on Control 1968.

APPENDIX A

REVIEW OF SYSTEM THEORY

A.0 SYSTEM THEORY

This appendix contains a review of the material in system theory, required in order to understand the formulation of dynamic stochastic systems.

A.1 MATHEMATICAL BACKGROUND

A.1.1 Vectors

A vector is a column of n elements. When $n = 1$ it is a *scalar*. A raw vector is a column vector transposed, of dimension $1 \times n$.

$$x^T = [x_1 \ \cdots \ x_n]. \tag{A.1}$$

Operations on a vector are addition, and multiplication by a constant,

$$x + y = [x_i + y_i], \qquad kx = [kx_i], \tag{A.2}$$

the inner product

$$x^T y = \sum_i (x_i y_i), \tag{A.3}$$

and the outer product or dyad,

$$xy^T = \{x_i y_j\}. \tag{A.4}$$

A metric of a vector is its length. The *Euclidean norm* is a metric given,

$$\|x\| = \sqrt{x^T x}\,. \tag{A.5}$$

The derivative of a vector and the integral of a vector are, respectively,

$$dx(t)/dt = [dx_1/dt \ \cdots \ dx_n/dt]^T, \qquad \int x(t)\,dt = \left[\int x_1\,dt \ \cdots \ \int x_n\,dt\right]^T, \tag{A.6}$$

assuming that the vector x is a function of time.

A.1.2 Matrices

A matrix is an $m \times n$ array with elements a_{ij}, $i = 1 \cdots m$, $j = 1 \cdots n$;

$$A = \{a_{ij}\}. \tag{A.7}$$

Operations on a matrix are addition, multiplication, and multiplication by a constant, respectively,

$$\begin{aligned} A + B = B + A &= \{a_{ij} + b_{ij}\}, \\ AB \neq BA &= \left\{\sum_k a_{ik} b_{kj}\right\}, \\ kA = Ak &= \{ka_{ij}\}, \end{aligned} \tag{A.8}$$

matrix differentiation and integration, respectively,

$$dA(t)/dt = \{da_{ij}/dt\}, \tag{A.9}$$

$$\int A(t)\,dt = \left\{\int a(t)_{ij}\,dt\right\}. \tag{A.10}$$

The identity matrix is

$$I = \{\delta_{ij}\}, \qquad \text{where } \delta_{ij} \text{ is the Kronecker delta } \delta_{ij} = \begin{cases} 1, & i = j, \\ 0, & i \neq j \end{cases}. \tag{A.11}$$

The determinant of a square $(n \times n)$ matrix is a scalar:

$$\det[A] = |A| = \sum_{i=1}^{n} \cdots \sum_{\substack{I=1 \\ I \neq i \cdots j}}^{n} [a_{1i} a_{2j} \ \cdots \ a_{nI}]. \tag{A.12}$$

A matrix with $\det[A] = 0$, is called *singular*.

The transpose of a matrix is

$$A^T = \{a_{ij}\}. \tag{A.13}$$

The adjoint of a matrix, Adj[A], is the matrix formed by replacing each element a_{ij} by the determinant of the submatrix formed by deleting the ith row and jth column of A and multiplying it by $(-1)^{i-j}$ (cofactor), and transposing the result.

A matrix inverse of a square matrix is defined to be

$$A^{-1} = |A|^{-1}\,\text{Adj}[A], \qquad AA^{-1} = I. \tag{A.14}$$

The trace of a matrix is the sum of the diagonal terms,

$$\text{Trace}[A] = \text{Tr}[A] = \sum_i a_{ii}. \tag{A.15}$$

Observe that

$$\text{Trace}[aa^T] = a^Ta.$$

If the inverse of a matrix does not exist, either because it is not square or its rank is less than maximum, it may be replaced by a pseudoinverse;

$$A^{\#} = (A^TA)^{-1}A^T. \tag{A.16}$$

Observe that $AA^{\#}A = A$ and $(A^{\#}A)^T = A^{\#}A$.

The characteristic equation of a matrix is

$$f(\lambda) = |\lambda I - A| = 0. \tag{A.17}$$

The solution of the characteristic equation yields the eigenvalues $\lambda = \lambda_1 \cdots \lambda_n$.

The rank of a matrix rank[A] is the size of its highest nonzero determinant of A.

A similarity transformation T on A, for which T^{-1} exists (nonsingular) yields a matrix B with the same eigenvalues;

$$TAT^{-1} = B. \tag{A.18}$$

The eigenvector of the $n \times n$ matrix A, e_i, associated with the eigenvalue λ_i, satisfies the equation

$$\lambda_i e_i = Ae_i. \tag{A.19}$$

The matrix $E = \{e_i\}$, composed of the eigenvectors of A, is nonsingular and satisfies the condition

$$AE = E\Lambda, \qquad \text{where } \Lambda = \text{diag}\{\lambda_i\}. \tag{A.20}$$

The Cayley–Hamilton Theorem states that replacing λ by A in the characteristic equation (A.17) yields

$$f(A) = 0. \tag{A.21}$$

This leads to the definition of the characteristic matrix,

$$e^A = I + A + \tfrac{1}{2}A^2 + \cdots. \tag{A.22}$$

A quadratic form is defined as

$$x^T A x = \sum_{ij} a_{ij} x_i y_j. \tag{A.23}$$

The norm of a matrix is defined as

$$\|A\| = \max_x \|Ax\|/\|x\| = \sqrt{\lambda_1}, \tag{A.24}$$

where λ_1 is the maximum eigenvalue of A.

Gradients of $z(x)$;

x scalar	$\partial z/\partial x = a$, scalar,	$\partial^2 z/\partial x^2 = A$, Hessian matrix;
x vector	$\partial z/\partial x = A$, matrix,	Jacobian;
X matrix	$\partial z/\partial X = B$, matrix.	

A.1.3 Vector Spaces

A *vector space* is a set whose elements obey the law of addition and multiplication by a constant. A *metric vector space* is one with a metric associated with it. The most common metric vector space is the Euclidean space, associated with the Euclidean norm (A.5).

A.1.4 Set Theory

A set is composed of elements with the same qualifier. The following operations: union ($\cup$), intersection ($\cap$), complement (overbar), and the infinite summation ($\bigcup_i^\infty$), null set ($\varnothing$), and whole space are defined as

$$A \cup B, \qquad A \cap B, \qquad \bar{A}, \qquad \bigcup_i^\infty A_i, \qquad \varnothing, \qquad S. \tag{A.25}$$

A σ field $\mathscr{F}$, is defined by sets $A \in \mathscr{F}$ for which $\bar{A} \in \mathscr{F}$. This implies that $\bigcup_i^n A \in \mathscr{F}$, and is closed under set operations. If $n = \infty$ it is a *Borel field*.

Theorem A.1 (DeMorgan)

$$\overline{(A + B)} = \bar{A}\bar{B}, \quad \text{and} \quad \overline{(AB)} = \bar{A} + \bar{B}. \tag{A.26}$$

Metrics $\mu(A)$ may be associated to vector spaces. Probability theory is a typical example of such a space, {S, $\mathscr{F}$, P}.

A.2 LINEAR SYSTEMS

A.2.1 Definitions

A state space is a Euclidean space with elements the states of a system. If $x(\cdot)$ are the states, $u(\cdot)$ the control inputs, and $w(\cdot)$ and $v(\cdot)$ the process and measurement noises, respectively, dynamic systems are classified as

(a) Continuous time, $x(t) \in X$;

$$dx(t)/dt = f(x, u, t, w), \qquad x(0) = x_0,$$
$$z(t) = h(x, u, t, v); \text{ or} \tag{A.27}$$

discrete time, $t = k\,\Delta t$, $x(k) = x(t) \in X_k$, $k = 1, 2, \ldots,$

$$x(k + 1) = f(x(k), u(k), k, w(k)), \qquad x(0) = x_0,$$
$$z(k) = h(x(k), u(k), k, v(k)); \tag{A.27A}$$

(b) Deterministic, if x_0, w, v, are deterministic quantities for all t or k, or stochastic, if x_0, w, v, are stochastic quantities for all t or k.

The equilibrium points of a dynamic system are defined by $dx/dt = 0$; for example, $f(x, u, \cdot\, , w) = 0$.

Linear systems are defined as having linear time-varying or time-invariant coefficients; for example,

$$dx/dt = F(t)x(t) + B(t)u(t) + G(t)w(t)x(0) = x_0,$$
$$z(t) = H(t)x(t) + v(t), \tag{A.28}$$

with $p(x_0)$, $p(w(t))$, $p(v(t))$ given for all $t \in [0, \infty]$.

Nonlinear systems may be linearized about a local (equilibrium) point. Canonic forms may be assigned to linear systems to use the minimum number of parameters. For a single-input–single-output system one identifies the following two canonic forms of equations.

State Canonic Form

$$F = \begin{bmatrix} 0 & I \\ & f^T \end{bmatrix}, \quad b^T = [0 \cdots 0, 1], \quad g^T = [0 \cdots 0, 1], \quad h^T = [h_1 \ \cdots \ h_n]. \tag{A.29}$$

Output Canonic Form

$$F = \begin{bmatrix} 0 & I \\ & f^T \end{bmatrix}, \quad b^T = [b_1 \ \cdots \ b_n], \quad g^T = [g_1 \ \cdots \ g_n], \quad h^T = [0 \cdots 0, 1]. \tag{A.30}$$

Similarity transformations T, Eq. (A.18), may transform a set of linear state equations to another, for example, a canonic form:

$$y(t) = Tx(t) \tag{A.31}$$

and

$$dy/dt = TFT^{-1}y(t) + TBu(t) + TGw(t) = Ay(t) + Ku(t) + Lw(t),$$

$$z(t) = HT^{-1}y(t) + v(t) = My(t) + v(t), \qquad y(0) = Tx(0), \tag{A.32}$$

where $TFT^{-1} = A$, $TB = K$, $TG = L$, and $HT^{-1} = M$.

In the case that $T = E = \{e_i\}$, the eigenvector matrix, $F = \Lambda = \text{diag}\{\lambda_i\}$, the diagonal eigenvector matrix, and the state equations are decoupled. The space $X \times T$, is called the solution space, and $\{x(t), t\} \in X \times T$. Similar results apply to discrete-time systems.

A.2.2 Solution of Linear Differential Equations

This approach applies to linear continuous-time systems of differential equations.

The State Transition Matrix, homogeneous solution. Define the state transition matrix as the homogeneous solution of the linear system Eq. (A.28);

$$dx/dt = F(t)x(t), \qquad x(t) = \Phi(t,0)C, \qquad C = \text{constant};$$

then

$$dx/dt = d\Phi/dt\, C = F\Phi(t,0)C, \qquad \forall\, C, \qquad \Rightarrow d\Phi/dt = F\Phi(t,0). \tag{A.33}$$

If $C = x_0$, then $\Phi(0,0) = I$, and

$$x(t) = \Phi(t,0)x_0. \tag{A.34}$$

Properties of the State Transition Matrix

$$\Phi(t_2, t_1) = \Phi(t_2, t_0)\Phi(t_0, t_1) \tag{A.35}$$

$$\Phi(t,t) = \Phi(t,t_0)\Phi(t_0,t) = I \qquad \Rightarrow \Phi(t,t_0)^{-1} = \Phi(t_0,t). \tag{A.36}$$

The inverse of the state transition matrix always exists, because

$$\det[\Phi(t,t_0)] \neq 0, \qquad \forall\, t \in [0,\infty]. \tag{A.37}$$

Expanding (A.35) in Taylor series, and from (A.22);

$$\begin{aligned} x(t) &= x(t_0) + dx(t_0)/dt(t - t_0) + \tfrac{1}{2} d^2x(t_0)/dx^2(t - t_0)^2 + \cdots \\ &= \left[I + F(t - t_0) + \tfrac{1}{2}F^2(t - t_0)^2 + \cdots\right]x(t_0) = e^{F(t-t_0)}x(t_0). \end{aligned} \tag{A.38}$$

The State Transition Matrix, Particular Solution. The method to be used to obtain the particular solution of the differential equations is called variation of parameters. The complete solution will be obtained by the superposition integral, applicable because the system is linear.

The complete differential equations are

$$\begin{aligned} dx/dt &= F(t)x(t) + B(t)u(t) + G(t)w(t)x(0) = x_0, \\ z(t) &= H(t)x(t) + v(t). \end{aligned} \tag{A.28}$$

Assume that in the homogeneous solution the constant C is replaced by a time-varying $C(t)$:

$$\begin{aligned} x(t) &= \Phi(t,t_0)C(t), \\ dx/dt &= d\Phi/dt\, C(t) + \Phi(t,t_0)\, dC/dt \\ &= F(t)\Phi(t,t_0)C(t) + \Phi(t,t_0)\, dC/dt. \end{aligned} \tag{A.39}$$

Considering the original equation,

$$\begin{aligned} dx/dt &= F(t)x(t) + B(t)u(t) + G(t)w(t) \\ &= F(t)\Phi(t,t_0)C(t) + B(t)u(t) \\ + G(t)w(t) &= F(t)\Phi(t,t_0)C(t) + \Phi(t,t_0)\, dC/dt, \end{aligned}$$

and eliminating the common terms yields

$$\Phi(t, t_0)\, dC/dt = B(t)u(t) + G(t)w(t) \quad \text{or}$$

$$dC/dt = \Phi(t, t_0)^{-1}[B(t)u(t) + G(t)w(t)]. \qquad \text{(A.40)}$$

Integrating Eq. (A.40), and considering that $\Phi(t, t_0)^{-1} = \Phi(t_0, t)$ and $C(t_0) = x_0$,

$$C(t) = x_0 + \int_{t_0}^{t} \Phi(t_0, \tau)[B(\tau)u(\tau) + G(\tau)w(\tau)]\, d\tau.$$

Substituting back in Eq. (A.39),

$$x(t) = \Phi(t, t_0)x_0 + \int_{t_0}^{t} \Phi(t, t_0)\Phi(t_0, \tau)[B(\tau)u(\tau) + G(\tau)w(\tau)]\, d\tau$$

$$x(t) = \Phi(t, t_0)x_0 + \int_{t_0}^{t} \Phi(t, \tau)[B(\tau)u(\tau) + G(\tau)w(\tau)]\, d\tau, \qquad \text{(A.41)}$$

and

$$z(t) = H(t)\Phi(t, t_0)x_0 + \int_{t_0}^{t} H(t)\Phi(t, \tau)[B(\tau)u(\tau) + G(\tau)w(\tau)]\, d\tau + v(t). \qquad \text{(A.42)}$$

A.2.3 Solution of Linear Difference Equations

There are two ways to generate a discrete-time system. One is by approximating the time derivatives in a differential equation by replacing the infinitesimal dt with the finite time interval Δt; the other by considering a sampled data system, with a sample and a hold, which provides values of a system at discrete-time instants, for example a digital computer. The latter represents an exact representation of the continuous-time system if the Nyquist theorem is satisfied.

Discrete-Time System Approximation of Differential Equations. Consider the continuous-time system of Eq. (A.28), and write the definition of the derivative vector,

$$ds/(t)/dt = \lim_{\Delta t \to 0} [x(t + \Delta t) - x(t)]/\Delta t$$

$$= F'(t)x(t) + B'(t)u(t) + G'(t)w(t)$$

$$z(t) = H(t)x(t) + v(t).$$

Removing the limit operation, and replacing $t = k\,\Delta t$ by k, Eq. (A.28) is approximated by

$$x(k+1) = [I + F'(k)\Delta t]x(k) + B'(k)\Delta t u(k) + G'(k)\Delta t w(k)$$
$$= F(k)x(k) + B(k)u(k) + G(k)w(k), \qquad x(0) = x_0, \tag{A.43}$$

$$z(k) = H(k)x(k) + v(k),$$

and the probability distributions $P(x_0)$, $P(w(k))$, and $P(v(k))$ are defined accordingly.

Sample Data Representation of the Continuous-Time System. This representation starts with the solution of the differential equations (A.41), over the interval $[t, t+\Delta t]$

$$x(t+\Delta t) = \Phi(t+\Delta t, t)x(t)$$
$$+ \int_t^{t+\Delta t} \Phi(t+\Delta t, \tau)[B(\tau)u(\tau) + G(\tau)w(\tau)]\,d\tau,$$
$$z(t) = H(t)x(t) + v(t). \tag{A.44}$$

Considering that $u(\tau) = u(k)$ and $w(t) = w(k)$ are constant over $[t, t+\Delta t]$ and can be pulled out of the integral, replacing $t = k\,\Delta t$ with k yields

$$x(k+1) = \Phi(k+1, k)x(k) + u(k)\int_t^{t+\Delta t} \Phi(t+\Delta t, \tau)B'(\tau)\,d\tau$$
$$+ w(k)\int_t^{t+\Delta t} \Phi(t+\Delta t, \tau)G'(\tau)\,d\tau$$
$$= F(k)x(k) + B(k)u(k) + G(k)w(k),$$
$$z(k) = H(k)x(k) + v(k). \tag{A.45}$$

It is obvious that the two expressions of the discrete-time system, Eqs. (A.43) and (A.45), are identical. Therefore, there is a single solution for both. In the case of constant coefficients F, B, G, and H, the solution is

$$z(k) = HF^{k+1}x_0 + \sum_{i=1}^{k+1} HF^{i-1}[Bu(k-i) + Gw(k-i)] + v(k). \tag{A.46}$$

A.3 STABILITY

Stability is one of the most important properties of systems, from the engineering point of view. There are two kinds of stability considered herein: Bounded-Input–Bounded-Output stability (BIBO) for systems with zero initial conditions, and stability of the equilibrium (SE) for systems with zero inputs.

A.3.1 BIBO Stability

For continuous-time systems with $x_0 = 0$, the condition for BIBO stability is

$$\int_0^{\infty} \|\Phi(T,t)B(t)\|\,dt < \infty. \tag{A.47}$$

For discrete-time invariant systems with $x_0 = 0$, the condition for BIBO stability is

$$\sum_{i=0}^{\infty} \|F^{i-1}B\| < \infty. \tag{A.48}$$

A.3.2 Stability of the Equilibrium

For *continuous-time systems* with $u(t) = 0,\ \forall\ t \in [0, \infty]$, the condition for stability of the equilibrium is

$$\text{Re}[\lambda_i(F(k))] \le 0, \qquad i = 1 \cdots n. \tag{A.49}$$

Asymptotic stability of the equilibrium is guaranteed if

$$\text{Re}[\lambda_i(F(k))] < 0, \qquad i = 1 \cdots n. \tag{A.50}$$

For *discrete-time systems* with $u(k) = 0,\ k = 1, 2 \cdots$, the condition for stability of the equilibrium is

$$|\lambda_i(F(k))| \le 1, \qquad i = 1 \cdots n. \tag{A.51}$$

Asymptotic stability of the equilibrium is guaranteed if

$$|\lambda_i(F(k))| < 0, \qquad i = 1 \cdots n. \tag{A.52}$$

A.4 CONTROLLABILITY

Controllability is a property relating the states of a system to its input. It is the same for continuous-time or discrete-time systems and the same conditions apply to both of them.

Definition A.1 A system is said to be controllable if there exists an input u that can take it from a point in the state space to another point, in finite time.

Definition A.2 A system is said to be completely controllable if there is an input u that can take it from *any* point in the state space to *any* other point, in finite time.

The condition of complete controllability for a time-invariant continuous-time or discrete-time system requires that the following matrix has maximum rank:

$$\text{rank}[B, FB, F^2B \ \cdots \ F^{n-1}B] = \text{max}. \tag{A.53}$$

A.5 OBSERVABILITY

Observability is the property of a continuous-time or discrete-time system to observe all the states of the system from its outputs z. The conditions for observability are the same for continuous-time or discrete-time systems.

Definition A.3 A system is said to be observable, at some time t_1 or $k\,\Delta t \in [0, T]$, if its states can be determined from its outputs $z(\cdot)$, measured over $[0, T]$.

Definition A.4 A system is said to be completely observable, at *all* times t_1 or $k\,\Delta t \in [0, T]$ if its states can be determined from its outputs $z(\cdot)$, measured over $[0, T]$.

The condition of complete observability for a time-invariant continuous-time or discrete-time system requires that the following matrix has maximum rank:

$$\text{rank}[H, F^TH, F^{T2}H \ \cdots \ F^{Tn-1}H] = \text{max}. \tag{A.54}$$

A.6 EQUIVALENCE OF FREQUENCY AND TIME DOMAIN

Frequency-domain approaches for describing linear time-invariant systems, by transforming them using Laplace transforms for continuous-time systems, or Z transforms for discrete-time systems, are rarely used in this text. Therefore, the discussion of frequency-domain techniques will be limited to the demonstration of the equivalence of the time-domain responses with the frequency-domain transfer functions.

A transfer matrix of a linear continuous-time system, assuming zero initial conditions $x_0 = 0$, and noises $w(t) = 0$, $v(t) = 0$, is obtained by taking the

Laplace transform of Eq. (A.28):

$$\pounds[dx/dt - Fx(t)] = \pounds[Bu(t)], \qquad Z(s) = \pounds[Hx(t)], \tag{A.55}$$

yielding the input-output relation,

$$Z(s) = H(sI - F)^{-1}BU(s). \tag{A.56}$$

Subsequently, the transfer matrix is given by

$$G(s) = H(sI - F)^{-1}B = \pounds[\Phi(t)]. \tag{A.57}$$

Equivalently the Z-transfer matrix is given by

$$G(z) = H(zI - F)^{-1}B = Z[HF^iB]. \tag{A.58}$$

REFERENCES

Brogan, W. L., *Modern Control Theory*, Quantum Publishers, New York, 1974.

Gelb, A. (Ed.), *Applied Optimal Estimation*, MIT Press, Cambridge MA, 1974.

Kuo, B. C., *Digital Control Systems*, (2nd ed.), Saunders College Publishing, Ft. Worth, TX, 1992.

APPENDIX B

REVIEW OF PROBABILITY THEORY

B AXIOMATIC PROBABILITY THEORY

The axiomatic probability theory was introduced by Kolmogorov (1933), and was the first attempt to formulate an analytic theory based on three fundamental axioms. This approach utilizes set and measure theories to map uncertain events in the space of their measure of uncertainty, and is thus able to perform known analytic operations for their study.

B.1 SET-THEORETIC DEFINITIONS

Set theory was presented in Appendix A, in conjunction with measure spaces $[S, \mathscr{F}, \mu]$, where S is a metric space, $\mathscr{F}$ is the σ-field of sets $A \subset S$, and $\mu(A)$ is a measure on all sets in S. This approach will be adapted to the space of experiments with uncertain outcome called probability space.

B.1.1 Definitions

Process is the class of problems under consideration.

Experiment is a particular run of the process.

Event is the occurrence of a phenomenon e, observed at a certain time.

Outcome is the result of an experiment, o.

Probability space is composed of outcomes of experiments.

Definition B.1 (set theoretic)

A space S with points e_i (events), outcomes o_i, a σ field $\mathscr{F}$ of sets $A \subset S$, and a probability measure $P(A)$ on all sets in S, is called a probability space $[S, \mathscr{F}, P]$ associated with an experiment Γ [Kolmogorov (1933)].

B.1.2 Probability Axioms

After establishing the set-theoretic concept of a probability space the following axioms on the probability measure $P(A)$, also due to Kolmogorov, are presented.

Probability Axioms

1. $0 \leq P(A)$.
2. $P(S) = 1$.
3. $P(A_1 \cup A_2 \cup \cdots \cup A_n) = \Sigma_i^n P(A_i)$, when A_i are mutually exclusive.
   (a) Axiom 3 applies for infinitely countable sets $n = \infty$.

The last axiom 3(a) is an extension of 3 to cover the whole real line. In the case that two sets A and B are not mutually exclusive, the following relation applies:

$$P(A \cup B) = P(A) + P(B) - P(A \cap B). \tag{B.2}$$

This is easily demonstrated by a Venn diagram.

B.1.4 Dependent Events and the Bayes Theorem

Conditional probabilities of *dependent events* are defined by using the chain rule:

$$P(A/B) = P(A \cap B)/P(B) = P(B/A)P(A)/P(B). \tag{B.3}$$

Theorem B.1 (Bayes) For events A_i, $i = 1 \cdots n$, that depend on a previous event B, the conditional probabilities are

$$P(A_i/B) = P(B/A_i)P(A_i)/\Sigma_j P(B/A_j)P(A_j). \tag{B.4}$$

If the events B and A_i, $i = 1 \cdots n$, are *statistically independent*, for example,

$$A \cap B = 0.$$

then [Papoulis (1991)]

$$P(A_i/B) = P(A_i), \qquad i = 1 \cdots n,$$
$$P(A_i \cap B) = P(A_i)P(B). \tag{B.5}$$

B.1.5 Repeated Trials

Combined experiments are the ordered pairs $(\zeta_1, \zeta_2) \in S, \zeta_1 \in S_1, \zeta_2 \in S_2$, where $S = S_1 \times S_2$ is the Cartesian product of the two sets of the original different experiments. In most cases the two experiments are assumed to be independent, for example, $P(\zeta_1 \times \zeta_2) = P(\zeta_1)P(\zeta_2)$. This can be extended to n sets. *Repeated trials* result in the Cartesian space $\mathfrak{C} = S \times S \times \cdots \times S$, of independent repetition of the same experiment. The most common class of such experiments are the repetitive tossing of a coin with possible outcomes heads h with probability p, or tails t with probability $q = 1 - p$, called Bernoulli trials. The probability that such an event occurs k times in n independent trials, studied by Bernoulli, is

$$P_n(k) = P(\text{h occurs } k \text{ times in any order}) = \binom{n}{k} p^k q^{n-k}; \qquad p + q = 1. \tag{B.6}$$

For k occurrences, $k_1 \leq k \leq k_2$,

$$P\{k_1 \leq k \leq k_2\} = \sum_{k=k_1}^{k_2} \binom{n}{k} p^k q^{n-k}. \tag{B.7}$$

B.1.6 Asymptotic Theorems

Theorem B.2 (DeMoivre–Laplace) If $npq \gg 1$, then

$$P_n(k) = \binom{n}{k} p^k q^{n-k} \approx 1/\sqrt{2\pi npq}\, \exp\left[-(k - np)^2/2npq\right]. \tag{B.8}$$

Using the above theorem, one may evaluate $P\{0 \leq k \leq k_2\}$,

$$\begin{aligned} P\{0 \leq k \leq k_2\} &= \sum_{k=0}^{k_2} \binom{n}{k} p^k q^{n-k} \\ &\approx 1/\sqrt{2\pi npq} \sum_{k=0}^{k_2} \exp\left[-(k - np)^2/2npq\right] \\ &\approx 1/\sqrt{2\pi npq} \int_{k=0}^{k_2} \exp\left[-(x - np)^2/2npq\right] dx \\ &= \operatorname{erf}\left\{[k_2 - np]/\sqrt{npq}\right\}. \end{aligned} \tag{B.9}$$

The *error function* is defined as

$$P(x) = \text{erf}\{x\} = 1/\sqrt{2\pi} \int_0^x \exp[-y^2/2]\, dy = \int_0^x g(y)\, dy, \quad \text{(B.10)}$$

and it has very interesting properties, for example,

$$\text{erf}(-x) = -\text{erf}(x),$$
$$\text{erf}(\infty) = \tfrac{1}{2}.$$

The density function $g(x)$ is known as the *Gaussian density function*, and is commonly used as a model of unknown noise in stochastic processes:

$$g(x) = 1/\sqrt{2\pi} \exp[-x^2/2]. \quad \text{(B.11)}$$

Theorem B.3 (the law of large numbers)

$$\lim_{n \to \infty} \text{Prob}\left\{ |1/n \sum_i^n x_i - p| \leq \epsilon \right\} = 1, \quad \text{(B.12)}$$

where p is the probability of success events $(x_i = 1)$.i

theorem was originally developed as a corollary of the DeMoivre–Laplace theorem.

Theorem B.4 (the strong law of large numbers)

$$\text{Prob}\left\{ \lim_{n \to \infty} |1/n \sum_i^n x_i - p| \leq \epsilon \right\} = 1. \quad \text{(B.13)}$$

This is a stronger version of the previous law due to Borel.

Finally, the central limit theorem, responsible for our preference to Gaussian distributions, establishes the asymptotic behavior of any random sequence to be normal.

Theorem B.5 (the central limit theorem)

$$E\{x_i\} = \eta_i, \qquad \sigma_{xi}^2 = \sigma_i^2 \quad \text{and} \quad x = \sum_i x_i, \qquad \eta = \sum_i \eta_i, \qquad \sigma^2 + \sum_i \sigma_i^2.$$

Then, regardless of the individual $p(x_i)$'s, the density of the average is

$$p(x) = \lim_{n \to \infty} \left\{ 1/\sigma\sqrt{2\pi} \exp\left[-(x - \eta)^2/2\sigma^2\right] \right\}. \quad \text{(B.14)}$$

Theorem B.6 (Poisson theorem) If $n \to \infty$, $p \to 0$, and $np \to \alpha$, then from Theorem B.2,

$$[n!/k!(n-k)!]\, p^k q^{n-k} \to e^{-\alpha}\alpha^k/k!. \tag{B.15}$$

An important application of Poisson's theorem is the approximate evaluation of n points randomly placed in the interval $(0, T)$;

$$P\{k \text{ points in } (t_1, t_2) \text{ interval}\} = \binom{n}{k} p^k q^{n-k},$$

when n is large and the interval $t_a = t_2 - t_1$ is small compared to T. Then

$$P\{k \in t_a\} \approx \exp[-nt_a/T](nt_a/T)^k/k!.$$

As $n \to \infty$, $T \to \infty$, $\lambda \to n/T < \infty$, and

$$P\{k \in t_a\} \approx \exp[-\lambda t_a](\lambda t_a)^k/k!, \tag{B.16}$$

which is known as the Poisson distribution.

B.2 RANDOM VARIABLES

B.2.1 Definitions

Definition B.2

Random variables (rv) are numbers $X(\zeta) \in R^1$, assigned to the outcomes ζ of an experiment Γ on S with $\mathscr{F}$ and $P(\zeta)$, and defined on the real line $R^1 = (-\infty, +\infty)$ [Papoulis (1991)]. The probability distributions $P(x)$ and densities $p(x)$ are assigned to the random variables, respectively, according to the rule, for continuous time,

$$P(x) = \text{Prob}(X \le x) = \int_0^x p(x)\, dx, \qquad \int_{-\infty}^{\infty} p(x)\, dx = 1,$$
$$p(x) = dP(x)/dx. \tag{B.17}$$

And for discrete time

$$\begin{aligned} P(x) &= \text{Prob}(X = X_i \le x) = \sum_i p_i u(x - x_i) \\ p(x) &= \lim_{n\to\infty} [P(x + dx) - P(x)]/dx \\ &= \lim_{n\to\infty} P(x \le X = X_i \le x + dx)/dx \\ &= \sum_i p_i\, \delta(x - x_i), \end{aligned} \tag{B.18}$$

where $p_i = p(\zeta_i)$, $u(x - x_i)$ is the step, and $\delta(x - x_i)$ the delta functions at x_i.

Properties of Distribution Functions

a. $P(-\infty) = 0,\ P(+\infty) = 1.$
b. $P(x_1) \le P(x_2)$, for $x_1 < x_2$.
c. $P(x)$ is right semicontinuous; for example, $P(x^+) = P(x)$.
d. $P(x_1 \le X \le x_2) = P(x_2) - P(x_1)$.

B.2.2 Joint Density Functions

Joint density functions of two or more random variables $X, Y \cdots$ are defined by

$$P(x, y \cdots) = \text{Prob}(X \le x \cap Y \le y \cap \cdots),$$

$$p(x, y \cdots) = \partial^2 (P(x, y) \cdots)/\partial x\, \partial y \cdots. \tag{B.19}$$

The marginal probabilities are then specified as

$$P(x \cdots) = P(x, \infty \cdots),$$

$$p(x \cdots) = \int_{-\infty}^{\infty} p(x, y \cdots)\, dy \cdots. \tag{B.20}$$

If X and Y are statistically independent, then

$$P(x, y) = P(x)P(y),$$

$$p(x, y) = p(x)p(y). \tag{B.21}$$

B.3 FUNCTIONS OF RANDOM VARIABLES

The concept of the function of a random variable is generated by a deterministic relation between random variables.

Definition B.3

If for a given outcome ζ of an experiment Γ, $X(\zeta)$ is a random variable, another random variable $Y(\zeta)$ is a function of $X(\zeta)$ with value

$$Y(\zeta) = g[X(\zeta)], \tag{B.22}$$

The probability distribution and density of the new random variable $Y(\zeta)$, are defined, respectively,

$$P_Y(y) = P[Y \leq y] = P[g(X) \leq y]$$
$$p_Y(y) = dP_Y(y)/dy. \qquad (B.23)$$

Because the function $g(x) = y$ may have n roots the following theorem determines the probability distribution and density function of Y.

Theorem B.6 If $x_1(y), x_2(y) \cdots x_n(y)$ are the real roots of (B.24), satisfying the relations $y = g(x_1) = g(x_2) = \cdots = g(x_n)$, then the density $p_Y(y)$ is

$$p_Y(y) = p_X(x_1)/|dg(x_1)/dx_1| + \cdots + p_X(x_n)/|dg(x_n)/dx_n|. \qquad (B.24)$$

The probability distribution is defined accordingly.

B.4 MOMENTS OF CONTINUOUS-TIME RANDOM VARIABLES

The nth moment of a continuous-time random variable is defined as

$$\bar{x}^n = E\{X^n\} = \int_{-\infty}^{\infty} x^n p_X(x)\, dx. \qquad (B.25)$$

Common moments are

$$\overline{X} = E\{X\} = \int_{-\infty}^{\infty} xp(x)\, dx \qquad \text{(mean value)},$$

$$\overline{X}^2 = E\{X^2\} = \int_{-\infty}^{\infty} x^2 p(x)\, dx \qquad \text{(second moment)}.$$

Central moments are defined as

$$E\{(X - \overline{X})^n\} = \int_{-\infty}^{\infty} (x - \overline{X})^n p_X(x)\, dx. \qquad (B.26)$$

where $n = 1, 2 \cdots$ the order of the central moment.

B.5 MOMENTS OF DISCRETE-TIME RANDOM VARIABLES

The nth moment of a discrete-time random variable is defined as

$$\overline{X}^n = E\{X^n\} = \lim_{N \to \infty} \sum_{i=1}^{N} X_i^n p_X(x_i). \qquad (B.27)$$

Common moments are

$$\bar{X} = E\{X\} = \lim_{N\to\infty} \sum_{i=1}^{N} X_i p_X(x_i), \tag{B.28}$$

$$\bar{X}^2 = E\{X^2\} = \lim_{N\to\infty} \sum_{i=1}^{N} X_i^2 p_X(x_i). \tag{B.29}$$

Central moments are defined as

$$E\{(X - \bar{X})^n\} = \lim_{N\to\infty} \sum_{i=1}^{N} (X_i - \bar{X})^n p_X(x)\, dx. \tag{B.30}$$

B.6 RANDOM SEQUENCES

The various definitions of stochastic convergence are given in Section 1.8.

B.6.1 The Chebyshev Inequality

This is a very useful inequality for the proof of convergence of stochastic sequences encountered in stochastic processes:

$$\text{Prob}\{\|x_n - x\| > \epsilon\} \le E\{\|x_n - x\|^2\}/\epsilon^2. \tag{B.31}$$

REFERENCES

Kolmogorov, A., "Grundbegriffe der Wahrscheinlichkeitsrechnung," *Ergeb. Mat. und ihrer Grenzg.*, Vol. 2, No. 3 (1933).

Papoulis, A., *Probability, Random Variables, and Stochastic Processes* (3rd ed.), McGraw-Hill, New York, 1991.

Parzen, E., *Stochastic Processes*, Holden-Day, Inc. San Francisco, CA, 1962.

INDEX